PRACTICAL RAMAN
SPECTROSCOPY –
AN INTRODUCTION

PRACTICAL RAMAN SPECTROSCOPY – AN INTRODUCTION

Peter Vandenabeele
Ghent University, Belgium

WILEY

This edition first published 2013
© 2013 John Wiley & Sons, Ltd

Registered office

John Wiley & Sons Ltd, The Atrium, Southern Gate, Chichester, West Sussex, PO19 8SQ, United
Kingdom

For details of our global editorial offices, for customer services and for information about how to apply
for permission to reuse the copyright material in this book please see our website at www.wiley.com.

Library of Congress Cataloging-in-Publication Data

Vandenabeele, Peter.
  Practical Raman spectroscopy : an introduction / Peter Vandenabeele, Ghent University, Belgium.
    pages cm
  Includes bibliographical references and index.
  ISBN 978-0-470-68319-4 (hardback) – ISBN 978-0-470-68318-7 (paperback) 1. Raman
spectroscopy–Study and teaching. I. Title.
  QD96.R34V36 2013
  543′.57–dc23

                                                                                        2013013049

A catalogue record for this book is available from the British Library.

Print ISBN: Cloth 9780470683194
            Paper 9780470683187

Set in 11/13pt Times by Laserwords Private Limited, Chennai, India

Printed and bound in Malaysia by Vivar Printing Sdn Bhd

2   2014

To my family – for all your support,
To my colleagues – that became friends,
To my students – who have never stopped inspiring me.

About the Author

Peter Vandenabeele obtained his masters' degree in chemistry at Ghent University, where he made his masters' thesis on thermal analysis of precursors for the synthesis of superconductors. His PhD research was carried out at the same university, but in the department of analytical chemistry, under the supervision of Prof. Dr. L. Moens. This research was on the optimisation of micro-Raman spectroscopy and total-reflection X-ray fluorescence for art analysis (2000). During his post-doctoral period Peter further worked on novel applications of Raman spectroscopy, in art analysis as well as in pharmaceutics, microbiology and astrobiology. In 2007, the author was appointed as research professor in the department of archaeology of Ghent University, where he further can apply his analytical skills to study archaeological and artistic objects.

Peter Vandenabeele has authored almost 100 research papers on Raman spectroscopy and in archaeometry. He has given many presentations on international conferences, of which several invited or plenary oral presentations. On many occasions he has written book chapters on Raman spectroscopy in archaeometrical research. As research professor, he has limited time to teach, but nevertheless, he enjoys introducing students in archaeometry as well as in Raman spectroscopy.

Contents

Preface

Raman spectroscopy is a very versatile molecular spectroscopic technique, with many different applications in a range of research fields. Whereas in its early days using this technique was very time-consuming and complex – and only applied in a few very specialised laboratories – today the technique is becoming increasingly popular in fundamental research as well as in applied science. Indeed, due to many instrumental evolutions, Raman spectroscopy has become increasingly more accessible and affordable. As a consequence, the technique has moved from the specialised laboratories towards more generally oriented laboratories. However, along with this broadening of applications, there is an increasing chance for misinterpretations and good training in Raman spectroscopy can help in avoiding these pitfalls.

This handbook starts with an introduction, where the history of Raman spectroscopy is sketched. In Chapter 1, the theoretical background of the technique is described. This theory is used to understand possible interferences (Chapter 2) and to study possible techniques to enhance the Raman intensity (Chapter 3). Chapter 4 focuses on the technical aspects of Raman spectroscopy: general aspects of Raman spectrometer construction and the properties of the different components. Together with these aspects, some considerations about noise in Raman spectra are discussed. The final chapter in this book (Chapter 5) describes aspects from daily practices in a Raman spectroscopy laboratory. Common approaches are described, such as smoothing operations, baseline corrections, spectral searches, etc. This chapter tries to explain a little about what lies behind the buttons in your spectroscopy software package.

Throughout the book the readers are exposed to questions, where the answers are listed at the end of the book. We also provide discussion topics throughout the text, where the reader can evaluate whether he/she understood everything all right. In this book, we also provide some intermezzos – short texts illustrating or providing an illustration or background information on a topic that is discussed in the main text. In the end of the book we also give a short literature list for the interested reader.

I hope that this book can help people who are new in the field of Raman spectroscopy to understand the approach and to avoid some common pitfalls. May this encourage people to further explore and develop the broad range of possibilities that this technique offers!

Peter Vandenabeele
Ghent University, Belgium

Acknowledgements

It would not have been possible for me to perform and develop my research in Raman spectroscopy without continuous support from many colleagues and friends – too many to name them all so I'll just mention three: Professor Luc Moens gave me the opportunity, freedom and support to explore different aspects of this technique; Professor Bernard Gilbert always encouraged me to continue this research and patiently introduced me to many aspects of Raman spectroscopy; finally, Professor Howell Edwards has continued to inspire me in many discussions on all sorts of Raman spectroscopy applications.

Writing a book is a demanding job, that took much more time than initially expected. All people at John Wiley were, however, still supportive and understanding. Thanks to all, especially Jenny Cossham, Sarah Tilley, Zoë Mills, Jasmine Kao, Krupa Muthu and Martin Noble.

Finally, I would like to thank my wife Isabel and my daughters for their support and understanding, especially on the evenings when I was writing behind my computer.

I hope that you, dear reader, enjoy reading through this book.

Acronyms, Abbreviations and Symbols

$\bar{v} = \dfrac{1}{\lambda}$	Absolute wavenumber
$\bar{\alpha}$	Average polarisability
%T	Percentage of the transmitted light
μ	Reduced mass of a molecule
c	The speed of light
CCD	Charge coupled device
E	Electrical field
F	Force
h	Planck's constant (h $= 6.6260755_{40} \cdot 10^{-34}$ J $\cdot$ s)
I	Intensity
k	Boltzmann constant (k $= 1.380658_{12} \cdot 10^{-23}$ J $\cdot$ K^{-1})
n	Refractive index
N.A.	Numerical aperture
O.D.	Optical density
p	Dipole moment
p_0	Amplitude of the oscillating dipole moment
q	Displacement
Q_k, Q_l	Normal coordinates, corresponding with the kth and lth normal vibration
Q_{v0}	Amplitude of the normal vibration

SD	Standard deviation of the signal
t	Time
U	Potential energy
v	Vibrational quantum number
α	Polarisability
α_{aniso}	Anisotropic polarisability tensor
α_{iso}	Isotropic polarisability tensor
β	Hyperpolarisability
γ	2nd hyperpolarisability
γ	Anisotropy factor
δ	Bending vibration
ε_0	Permittivity of vacuum $(8.854187817 \cdot 10^{-12}\,C^2 \cdot N^{-1} \cdot m^{-2})$
κ	Force constant of a bond
ν	Stretching vibration
ρ	Degree of depolarisation
ρ	Rotation
σ	Noise
Φ	Electromagnetic flux
φ_v	Phase angle
Ω	Solid angle
ν_0	Frequency of incident radiation
ν_m	Frequency of measured radiation
ν_o	Vibrational frequency of the electromagnetic radiation
ν_v	Vibrational frequency of the molecule
ω	Raman wavenumber (in cm^{-1})
ψ_v	Vibrational wave function

Introduction to Raman Spectroscopy

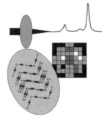

I'm picking up good vibrations
She's giving me the excitations
Good, (bop bop) good vibrations

The Beach Boys, 'Good Vibrations', 1966

Learning Objectives

- To appreciate the historical background of Raman spectroscopy
- To understand how instrumental improvements opened the way to new Raman spectroscopic applications

It was on 28 February 1928 that Sir C.V. Raman and K.S. Krishnan for the first time succeeded in demonstrating the inelastic scattering of light by a fluid. For this work, in 1930, Raman was honoured with the Nobel Prize. At the time, they used filtered sunlight to excite the molecules and photographic plates were used to record the spectrum. It took about 24 hours to record a spectrum of a beaker with ca. 600 ml of pure liquid. Knowing this, it is clear that Raman spectroscopy for a long time was only limited to specialised research laboratories and that the technique was considered as a curiosum. Today, Raman spectroscopy has moved out of the highly specialised laboratories and is available not only in many research institutions, but also as a reliable technique for quality control or even for sorting out plastics in the recycling industry.

Indeed, in today's research, Raman spectroscopy is appreciated for many different reasons. First, the technique is a relatively fast method and well-suited to investigating solids, liquids, solutions and even gases, depending on the experimental set-up used. For the analysis of solids often barely – if any – sample preparation is needed: just position it under the microscope and focus the laser beam. Liquids can be measured through glass vials and, as opposed to infrared spectroscopy, the presence of water does not hamper the measurements. Small portable spectrometers are now available and fibre optics probe heads allow us to record spectra from a distance, which is useful for analysis in harsh conditions or for the investigation of, for instance, explosives. Mobile analysis allows objects to be in situ and in a noninvasive way. Micro-Raman spectroscopy is one of the rare spectroscopic methods that enables us to obtain molecular information at the micrometer-scale. Chemometrics can be used during Raman spectroscopy studies, and molecules are easily interpreted by using automated algorithms for searching spectral libraries.

Current Raman spectroscopy research is very different from the approach adopted in the early days. Instrumentation has seriously evolved, and as a consequence sample sizes and measuring times are seriously reduced. Historically it has been seen that (r)evolutions in Raman spectroscopy research (e.g. new applications, access for a larger group of scientists, etc.) are often caused by the availability of new equipment or instrumental innovations. A first improvement since Raman's days was the introduction of mercury lamps as a source of excitation. However, measuring procedures and alignment remained quite complex and Raman spectroscopy was for a long period only used in restricted research areas, while infrared spectroscopy gained in importance. A big step forward in Raman research was in the 1960s, with the advent of the first lasers. The instrumentation was still quite expensive and aligning the sample and setting up of experiments was not straightforward. Often double or triple monochromators were used. At the end of the 1980s, new progress was caused by new instrumentation, such as the introduction of CCD (charge-coupled device) detectors, holographic filters, new optics (micro-Raman spectroscopy and fibre-optics) and the introduction of FT-Raman spectroscopy (Fourier-Transform Raman spectroscopy). Moreover, price reductions and the miniaturisation of instruments meant that Raman spectroscopy was no longer limited to a few specialised research laboratories, and the technique has become increasingly more common in fundamental research as well as in industry.

Throughout this book, several common applications of Raman spectroscopy are given. Many more can be found in the literature. In the following chapters we want to introduce you to this approach, firstly by introducing the fundamentals of the technique and thereafter by discussing some interferences and possibilities to enhance the Raman signal. In a later chapter we will discuss the main components of current Raman spectroscopy instrumentation, while in the final chapter we try indicate to the reader some of the common pitfalls and approaches when using this technique.

Summary

In this introductory chapter, the reader has been introduced to some historical aspects of Raman spectroscopy. By now, it should be clear that in the past, many advances and new applications were caused by improvements in Raman spectroscopy instrumentation, such as the introduction of lasers or sensitive detectors.

Further Reading

Lewis, I.R. and Edwards, H.H.M. (eds), *Handbook of Raman Spectroscopy –
    From the Research Laboratory to the Process Line*, Marcel Dekker, Inc., New
    York, 2001.
McCreery, R.L., *Raman Spectroscopy for Chemical Analysis*, John Wiley, New
    York, 2000.

Chapter 1
Theoretical Aspects

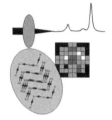

Learning Objectives

- To appreciate the historical background of Raman spectroscopy
- To understand how instrumental improvements opened the way to new Raman spectroscopic applications

The theoretical background of the Raman effect is already extensively described in literature. The Raman effect can be considered as the inelastic scattering of electromagnetic radiation. During this interaction, energy is transferred between the photons and the molecular vibrations. Therefore, the scattered photons have a different energy to the incoming photons.

1.1 Classical Approach

When a molecule is positioned in an electrical field $\mathbf{E}$, an electrical dipole moment $\mathbf{p}$ is induced. The relation between this induced

Practical Raman Spectroscopy: An Introduction, First Edition. Peter Vandenabeele.
© 2013 John Wiley & Sons, Ltd. Published 2013 by John Wiley & Sons, Ltd.

dipole moment and the electrical field can be expressed as a power series:

$$\mathbf{p} = \alpha \cdot \mathbf{E} + (1/2) \cdot \beta \cdot \mathbf{E}^2 + (1/6) \cdot \gamma \cdot \mathbf{E}^3 \ldots \qquad (1.1)$$

In this equation, α, β and γ are tensors, which are named polarisability, hyperpolarisability and 2nd hyperpolarisability, respectively. Typically, they are in the range of ca. $\alpha \sim 10^{-40}$ C $\cdot$ V$^{-1} \cdot$ m^2, $\beta \sim 10^{-50}$ C$\cdot$V$^{-2} \cdot$ m^3 and $\gamma \sim 10^{-60}$ C $\cdot$ V$^{-3} \cdot$ m^4. As these tensors each are a factor 10 billion less intense, the influence of these factors can in many cases be neglected.

QUESTION 1.1

Estimate value of the terms in Equation (1.1), when using an electrical field of $3 \cdot 10^6$ Vm^{-1} (corresponding to a typical laser intensity of ca. 10^9 Wm^{-2}).

The induced dipole moment can be thus considered as directly proportional to the electrical field and Equation (1.1) is reduced to:

$$\mathbf{p} = \alpha \cdot \mathbf{E} \qquad (1.2)$$

When studying the Raman effect, the electrical field is caused by electromagnetic radiation. Indeed, light can be considered as an oscillating electrical field. The electrical field vector $\mathbf{E}$ on the moment t is described as:

$$\mathbf{E} = \mathbf{E}_0 \cdot \cos(2\pi \cdot v_o \cdot t) \qquad (1.3)$$

with v_o the vibrational frequency of the electromagnetic radiation.

In Equation (1.2) the polarisability α is a tensor, which is dependent on the shape and dimensions of the chemical bond. As chemical bonds change during vibrations, the polarisability is dependent on the vibrations of the molecule. It can be said that the polarisability tensor (α) is dependent on the normal coordinate Q of the molecule. This relationship can be expressed as a Taylor series:

$$\alpha = \alpha_0 + \sum_k \left(\frac{\partial \alpha}{\partial Q_k}\right)_0 \cdot Q_k + \frac{1}{2}\sum_{k,l} \left(\frac{\partial^2 \alpha}{\partial Q_k \partial Q_l}\right)_0 \cdot Q_k \cdot Q_l + \ldots$$

(1.4)

Q_k and Q_l are the normal coordinates that correspond with the kth and l^{th} normal vibration, corresponding with the vibrational frequencies v_k and v_l. In a first approximation, only the first two terms in this equation are maintained. This means that the different (normal) vibrations are considered as totally independent and no cross-terms are included in the equation. Thus, considering the vth normal vibration, Equation (1.4) is reduced to:

$$\alpha_v = \alpha_0 + \alpha_v' \cdot Q_v$$

(1.5)

with α_v' the derivative of the polarisability tensor to the normal coordinate Q_v, under equilibrium conditions.

In a first approximation, the normal coordinate oscillates according to the harmonic oscillator (Intermezzo 1.1). The normal coordinate varies as a function of time according to:

$$Q_v = Q_{v0} \cdot \cos(2\pi \cdot v_v \cdot t + \varphi_v)$$

(1.6)

with Q_{v0} the amplitude of the normal vibration and φ_v a phase angle. Substitution of Equation (1.6) for Equation (1.5) gives:

$$\alpha_v = \alpha_0 + \alpha_v' \cdot Q_{v0} \cdot \cos(2\pi \cdot v_v \cdot t + \varphi_v)$$

(1.7)

When considering only the first two terms of the Taylor series (1.4), we assume that the polarisability tensor undergoes a harmonic oscillation, with a frequency v_v, that equals the vibrational frequency of the normal coordinate of the molecule. By substituting

Equations (1.7) and (1.3) in the (simplified) definition of the dipole moment **p** (Equation (1.2)), we obtain:

$$\mathbf{p} = \alpha_0 \cdot \mathbf{E}_0 \cdot \cos(2\pi \; \nu_o \; t) + \alpha'_v \cdot \mathbf{E}_0 \cdot Q_{v0} \; \cdot \cos(2\pi \; \nu_0 \; t)$$
$$\cdot \cos(2\pi \; \nu_v \; t \; + \varphi_v) \tag{1.8}$$

By using the trigonometrical formula:

$$\cos A \cdot \cos B = \tfrac{1}{2} \left[\cos(A+B) + \cos(A-B) \right] \tag{1.9}$$

Equation (1.8) can be modified to:

$$\mathbf{p} = \alpha_0 \cdot \mathbf{E}_0 \cdot \cos(2\pi \cdot \nu_o \cdot t)$$
$$+ \; \tfrac{1}{2} \cdot \alpha'_v \cdot \mathbf{E}_0 \cdot Q_{v0} \cdot \cos[2\pi \cdot (\nu_0 + \nu_v) \cdot t + \varphi_v]$$
$$+ \; \tfrac{1}{2} \cdot \alpha'_v \cdot \mathbf{E}_0 \cdot Q_{v0} \cdot \cos[2\pi \cdot (\nu_0 - \nu_v) \cdot t - \varphi_v] \tag{1.10}$$

Therefore, we can consider the induced dipole moment as a function of the vibrational frequencies of the molecule (ν_v) and of the incident radiation (ν_0):

$$\mathbf{p} = \mathbf{p}(\nu_0) + \mathbf{p}(\nu_0 + \nu_v) + \mathbf{p}(\nu_0 - \nu_v) \tag{1.11}$$

The induced dipole moment can be split into 3 components, each with a different frequency-dependence. The first term in Equation (1.11) corresponds to the elastic scattering of the electromagnetic radiation: the induced dipole moment has the same frequency (hence the same energy) as the incoming radiation. This type of scattering is called 'Rayleigh scattering', named after Lord Rayleigh (1842–1919), who used Rayleigh scattering to determine the size of a molecule. The 2nd and 3rd term in Equation (1.11) correspond to the inelastic scattering of light: Raman scattering. The 2nd term corresponds to a higher energy of the scattered radiation, compared to the incident beam (Anti-Stokes scattering), while the last term represents a lowering of the frequency (Stokes scattering).

When using Raman spectroscopy, an intense, monochromatic beam of electromagnetic radiation (usually a laser) is focussed on the sample, and the intensity of the scattered radiation is measured as a function of its wavelength. Usually, in a Raman spectrum the intensity is plotted as a function of the Raman wavenumber ω, expressed in cm^{-1}, which is related to the difference in frequency between the scattered light and the incident electromagnetic radiation:

$$\omega = \overline{\nu_m} - \overline{\nu_0} = \frac{\nu_m}{c} - \frac{\nu_0}{c} \qquad (1.12)$$

In this expression, the symbols ν_m and ν_0 stand for the frequency of the scattered (measured) and incident radiation, respectively; c is the speed of light; wavenumbers are usually expressed in cm^{-1}, so

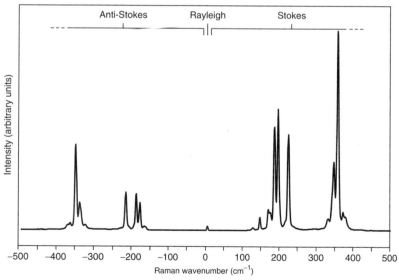

Figure 1.1 Raman spectrum of the mineral realgar (As$_4$S$_4$) in the anti-Stokes region, Rayleigh line and Stokes region. The actual intensity of the Rayleigh line is suppressed by a holographic filter in the spectrometer.

be careful when calculating them, as you need to use the appropriate units. Positive wavenumbers correspond with Stokes scattering, while negative wavenumbers correspond with anti-stokes scattering (see Figure 1.1).

QUESTION 1.2

The Raman spectrum in Spectrum 1.1 was recorded by using a laser with a wavelength of 785 nm. The most intense Raman band of realgar is positioned at $352\,\text{cm}^{-1}$. What are the (absolute) frequencies and wavelengths of the scattered radiation, in the Stokes as well as in the anti-Stokes region?

INTERMEZZO 1.1 THE HARMONIC OSCILLATOR AND THE POSITION OF THE VIBRATIONAL ENERGY LEVELS

When discussing molecular vibrations, bonds are often considered as a harmonic oscillator: a diatomic molecule is represented as two balls connected with a spring. For this oscillator Hooke's law is valid: $\mathbf{F} = -k \cdot q$ (with $\mathbf{F}$ a force which is in the opposite sense to the displacement $\mathbf{q}$, and with k the spring constant). In general, the force of the spring can be described as the (partial) derivative of the energy in function of the displacement: $-\mathbf{F} = \partial U/\partial q$. Therefore, the potential energy for a specific displacement q is given by the following Taylor-series:

$$U_q = U_{q=0} + (dU/dq)_{q=0} \cdot q + \tfrac{1}{2} \cdot (d^2U/dq^2) \cdot q^2 + \dots$$

For small displacements (i.e. the harmonic oscillator approximation), only the first two terms of this series are considered, and thus: $U = C + \tfrac{1}{2} \cdot k \cdot q^2$. (In general, the constant term C is set to 0, the energy state in equilibrium position.) In the case of the harmonic oscillator, a parabolic curve is obtained when plotting the potential energy as a function of the displacement.

When substituting this potential energy curve in the Schrödinger equation $\left(-\frac{h^2}{8\pi^2\mu} \cdot \frac{d^2\Psi}{dq^2} + U \cdot \Psi = E \cdot \Psi\right)$, one can determine the possible energy levels of the harmonic oscillator: $E_v = (v + 1/2) \cdot \frac{h}{2\pi}\sqrt{\frac{k}{\mu}}$, with the vibrational quantumnumber $v = 0, 1, 2, \ldots, \infty$. μ is the reduced mass of the molecule $\left(\frac{1}{\mu} = \frac{1}{m_1} + \frac{1}{m_2}\right)$. It can be noted that for the harmonic oscillator approximation, the vibrational energy levels are equidistant (see Figure Intermezzo 1.1).

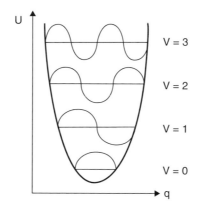

Figure Intermezzo 1.1 Harmonic oscillator.

1.2 Selection Rule

From Equation (1.10) it can be seen that the Raman effect only happens if $\alpha'_v \neq 0$. Therefore, during the considered normal vibration a change in polarisability should happen:

$$\alpha'_v = \left(\frac{\partial\alpha}{\partial Q_v}\right)_0 \neq 0 \qquad (1.13)$$

Opposite to this, in infrared spectroscopy there is a change in (permanent) dipole moment during the vibration. In principle, on ground of this selection rule, it can be determined whether a vibration is Raman or infrared active. However, only for very simple molecules this can be done by hand: vibrations of more complex molecules can be evaluated by using quantum mechanics and group theory.

QUESTION 1.3

Describe the different vibrations in a CO_2 molecule. Which of these vibrations is infrared active and which is Raman active?

Consider a homonuclear diatomic molecule (e.g. H_2). In this case, the polarisability is a function of the internuclear distance. This can be understood by thinking that, when the internuclear distance is larger, the electrons are relatively further away from the nuclei and thus can be moved more easily. During this vibration, a change in polarisability of the molecule occurs and thus the vibration is Raman active. The dipole moment does not change and the vibration is infrared inactive.

Analogously, in a heteronuclear diatomic molecule (e.g. HCl) polarisability is also dependent on the bond distance. Moreover, since the size of the dipole moment is dependent on the charge separation, during a vibration there is a change of the dipole moment. The vibration is thus Raman and infrared active.

For triatomic molecules, different cases can be distinguished, depending on the number of different types of atoms (1;2 or 3) and depending on the linearity of the molecule. For a linear molecule of the type ABA (e.g. CO_2), 4 normal vibrations can be distinguished: symmetrical stretch vibration, asymmetrical stretch and 2 bending vibrations (degenerated). Of these, only the symmetrical stretch is Raman active, whereas the other vibrations are infrared active.

For nonlinear molecules of the type ABA (e.g. H_2O), 3 normal vibrations can be distinguished (symmetrical and asymmetrical stretch, symmetrical bending vibration) that are infrared and Raman active. In general it can be stated that, in a centrosymmetric molecule Raman active vibrations are infrared inactive and vice versa. Moreover, from practice we know that vibrations that give rise to very intense infrared bands are generally rather weak Raman scatterers.

To determine the Raman and/or infrared activity of vibrations in complex molecules, quantum mechanics and group theory are used. The total vibrational wave function ψ_v can be written as the product of different wave functions ψ_i (n_i), with ψ_i the ith normal vibration that is currently in the n_ith state:

$$\psi_v = \psi_1(n_1) \cdot \psi_2(n_2) \cdot \ldots \cdot \psi_k(n_k) = \prod_{j=1}^{k} \psi_i(n_i) \qquad (1.14)$$

A fundamental transition is the transition form a state with all normal vibrations in the ground state $\psi_i(0)$ to a state with all vibrations but one in the ground state. That particular vibration is in the first excited state ψ_i *(1)*:

$$\prod_i \psi_i(0) \rightarrow \psi_j(1) \cdot \prod_{i \neq j} \psi_i(0) \qquad (1.15a)$$

This vibrational transition can be concisely written as:

$$\psi_v^0 \rightarrow \psi_v^{j=1} \qquad (1.15b)$$

The induced dipole moment is given by Equation (1.2). The amplitude of the transition moment which is induced during transition 1.15, is expressed as:

$$[p_0]_{j=1 \leftarrow 0} = \int \psi_v^{j=1*} \, \widehat{\alpha} \psi_v^0 d\tau \cdot \mathbf{E}_0 \qquad (1.16)$$

   The operator $\widehat{\alpha}$ is the polarisability (cf. Equation (1.1)) and τ is the spatial coordinate. Here, we only consider the time-independent part (i.e. the amplitude) of the transition moment, which allows us to use the time-independent part of the perturbation theory. In analogy with Equation (1.2), we can consider the integral in Equation (1.16) as the quantum mechanical equivalent of the polarisability tensor. This integral can in a simply way be noted in the Dirac notation:

$$[\alpha]_{j=1\leftarrow 0} = \int \psi_v^{j=1*} \widehat{\alpha} \psi_v^0 d\tau = \langle \psi_v^{j=1} | \alpha | \psi_v^0 \rangle \qquad (1.17)$$

   Polarisability is a function of the normal coordinate. We assume harmonicity, and thus we only substitute the first two terms of Equation (1.4), leading to:

$$[\alpha]_{j\leftarrow 0} = \alpha_0 \cdot \langle \psi_v^j | \psi_v^0 \rangle + \sum_k \left(\frac{\partial \alpha}{\partial Q_k} \right)_0 \cdot \langle \psi_v^j | Q_k | \psi_v^0 \rangle \qquad (1.18)$$

   From this equation, the quantum mechanical description of the border conditions of the Raman effect can be derived. Because of the orthogonality of the wave functions, the value of the integral in the first term is:

$$\langle \psi_v^j | \psi_v^0 \rangle = 0 \text{ if } j \neq 0 \qquad (1.19a)$$

$$\langle \psi_v^j | \psi_v^0 \rangle = 1 \text{ if } j = 0 \qquad (1.19b)$$

   This term corresponds to Rayleigh scattering. The second part of Equation (1.18) describes the Raman scattering. The condition for the Raman effect to happen is that both factors in this term differ from 0. The first factor indicates that the polarisability should change during the vibration, whereas the second factor only differs from 0 if the wave function ψ_v^j and the normal coordinate Q_k are determined by the same quantumnumber ($j = k$). When considering

Stokes Raman scattering, the transition from the ground state to the first excited vibrational state ($j = 1$) is involved.

$$\langle \psi_v^{j=1} | Q_k | \psi_v^0 \rangle \neq 0 \quad \text{if} \quad j = k = 1 \tag{1.20}$$

As a consequence, for Stokes scattering, Equation (1.18) should be written as:

$$[\alpha]_{j=1 \leftarrow 0} = \left(\frac{\partial \alpha}{\partial Q_k} \right)_0 \cdot \langle \psi_v^1 | Q_k | \psi_v^0 \rangle \tag{1.21}$$

The selection rule states that the polarisability for the considered transition (Stokes Raman) should differ from 0. To evaluate this, it is not necessary to calculate the integral as such; based on its symmetry properties it can be determined whether it is different from 0 or not. This is only the case if ψ_v^j and Q_k belong to a different symmetry class, which means that all possible symmetry operations are transformed in the same way. Based on the symmetry properties of a molecule, it can be determined how a certain vibration transforms, how its symmetry class can be determined, and whether a vibration is Raman active or not. These studies are the subject of group theoretical examinations. When considering (crystalline) solids, not only the symmetry of the 'molecule' or the ion (e.g. CO_3^{2-}) should be taken into account, but also the geometry in the crystalline structure. Symmetry of the cavity that contains the group plays an important role.

INTERMEZZO 1.2 THE ANHARMONIC OSCILLATOR AND THE POSITION OF VIBRATIONAL ENERGY LEVELS

In this discussion, molecular vibrations are considered as harmonic oscillations. In reality, however, this is not the case and harmonic oscillations are a simplified model. If molecular vibrations were harmonic oscillations, atoms could approach each other until they overlap entirely – which obviously is impossible. Moreover, atoms would always oscillate around their equilibrium distance, also if they had been taken a long distance apart. This is not the case as molecular bonds break when the interatomic distance is too large.

Therefore, there are deviations of the model of the harmonic oscillator. The curve describing the potential energy of the anharmonic oscillator (see Figure Intermezzo 1.2) as a function of the distance between the atoms' nuclei, is named the Morse curve, with the equation $U = D_e.(1 - e^{-\beta q})^2$. The Morse curve is presented in the illustration below. It can be seen that atoms can approach each other less easily, compared to the harmonic oscillator. Moreover, if atoms are sufficiently far apart (i.e. when they are no longer bound to each other), there is no change in the potential energy curve.

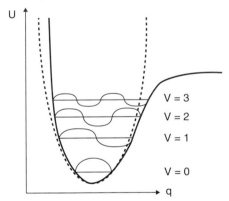

Figure Intermezzo 1.2 Anharmonic oscillator.

A consequence of the introduction of the anharmonic oscillator is that the distance between consecutive energy levels is not equal. In spectroscopy, this is seen by the occurrence of sum peaks and overtones.

1.3 Energy Levels and Group Frequencies

The result of the Schrödinger equation for a particle in a box tells us that only a limited number of energy levels are allowed:

$$E_\nu = (\nu + ^1\!/_2) \cdot h\nu_\nu \quad (\nu = 0, 1, 2, \dots) \qquad (1.22)$$

With ν the vibrational quantumnumber, h Planck's constant ($h = 6.6260755_{40} \cdot 10^{-34}$ J·s) and ν_ν the molecular vibrational frequency. During the transition of one vibrational level to the next, the difference in quantumnumbers between begin and end state should equal 0 or ± 1. This is a consequence of the use of the harmonic oscillation approximation.[1]

If the difference equals 0, Rayleigh scattering is considered, otherwise Stokes or anti-Stokes (Raman) scattering is discussed. It is thus clear that the *overall* energy difference for a Raman transition equals $\pm h \cdot \nu_\nu$. This energy difference is plotted on the horizontal axis of a Raman spectrum, expressed as Raman wavenumbers (Equation (1.12)).

The position of a Raman band in the spectrum is determined by the energy difference between the ground state and the first vibrationally excited state (Figure 1.2).

As a consequence of the harmonic oscillation approximation, the vibrational frequency can be expressed as:

$$\nu_\nu = \frac{1}{2 \cdot \pi} \cdot \sqrt{\frac{\kappa}{\mu}} \qquad (1.23)$$

with κ the force constant of the bond and μ the reduced mass. For a diatomic molecule, the reduced mass is defined as:

$$\frac{1}{\mu} = \frac{1}{m_1} + \frac{1}{m_2} \qquad (1.24)$$

---

[1] This approach is usually allowed for traditional Raman spectroscopic experiments. Overtones and sumpeaks (that do not obey this approximation) are usually weak.

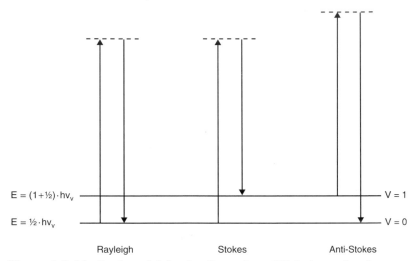

$E = (1+\tfrac{1}{2})\cdot hv_v$ V = 1

$E = \tfrac{1}{2}\cdot hv_v$ V = 0

Rayleigh   Stokes   Anti-Stokes

Figure 1.2 Idealised model for the dispersion of light by molecules.

with m_1 and m_2 the masses of the atoms constituting the bond. For more complex systems, where multiple atoms are involved, this equation can easily be expanded.

Equation (1.23) forms the basis for the interpretation of Raman spectra. From this equation, it appears that two factors determine the Raman band position: the force constant of the bond and the type of vibration (κ) and the reduced mass (μ). Once these two properties are known, the Raman band position can be calculated. Opposite, one can calculate the force constant of a bond from the Raman band position (provided the (reduced) mass is known). Since κ and μ are relatively stable for a specific type of bond or a particular functional group, for a series of functional groups so-called group frequencies can be determined. These characteristic wavenumbers give an indication about the region where a particular Raman band can be expected, but don't say whether this band is present or not, as this is dependent on the selection rules (§ 1.2) and the final appearance of the spectrum is strongly dependent on the symmetry of the molecule. A typical example of this is the Raman spectrum of substituted benzene molecules, where the appearance of the spectrum not only is determined by the substituents, but mainly by the number of substituents and their relative positions on the benzene ring (symmetry see §5.3.3).

The idea of group frequencies is based on the simplification that vibrations in a specific functional group are independent of the other vibrations in the molecule. These group frequencies have mainly practical value for the interpretation of Raman spectra of organic molecules. Often Raman spectra of inorganic molecules are more influenced by the symmetry properties of the molecule. Moreover, strong band shifts occur frequently as well as possible overlap between functional groups. For instance, there is overlap between the regions of total symmetrical stretch vibration of the carbonate (CO_3^{2-}) and sulphate (SO_4^{2-}) ions. In Table 1.1 some group frequencies of organic molecules are mentioned.

Note that stretching vibrations are usually abbreviated with the Greek letter nu (ν), while bending vibrations have the symbol delta (δ) and vibrations involving rotations are assigned the Greek letter rho (ρ).

Discussion Topic 1.1

In Table 1.1 some general trends can be observed:

- vibrations where H-atoms are involved, usually occur at higher wavenumbers;
- ν(C–C) have lower group frequencies than ν(C=C) and than ν(C≡C);
- stretching vibrations have higher group frequencies than the corresponding bending vibrations.

Explain why.

Answer:

From Equation (1.24) it can be seen that molecules involving light atoms (like hydrogen) have a smaller reduced mass than molecules with large atoms. As a consequence their vibrational frequency is higher (Equation 1.23). Opposite to this, inorganic materials, that often contain heavy atoms or ions, usually have Raman bands at lower wavenumbers. This property can be used to distinguish between spectra of inorganic or organic materials.

From Equation (1.23) it can be seen that the force constant κ is the second factor that influences the Raman band position. Bonds with high force constants (more or less synonym of 'strong bonds') can give rise to bands at high wavenumber positions. This can be seen in the group frequences of the series $v(C\equiv C)$, $v(C=C)$, and $v(C-C)$.

When evaluating Table 1.1 we also see that stretch vibrations occur at high wavenumbers, compared to bending vibrations. During stretch vibrations the bond length changes, while during bending vibrations only the bond angle is changed. Since another type of vibration is involved, another force constant should be used. Changing a bond length requires more energy than bending it, hence the difference in Raman band positions.

The force constant is strongly dependent on inter- and intramolecular interactions. For instance, hydrogen bonding can cause a band shift in some molecules. Substituents on a functional group can as well cause a change in the force constant κ. Electron donors or electron acceptors influence the electron density of the bond, and thus also its force constant, which is observed as a shifting Raman band.

Table 1.1 Some group frequencies for functional groups in organic compounds.

Vibration		Wavenumbers (cm^{-1})	Raman-intensity[1]
O–H stretch	ν(O–H)	3650–3000	w
N–H stretch	ν(N–H)	3500–3300	m
C–H stretch of alkynes	ν($\equiv$C–H)	3350–3300	w
C–H stretch of alkenes	ν(= C–H)	3100–3000	s
C–H stretch of alkanes	ν(–C–H)	3000–2750	s
C$\equiv$C stretch of alkynes	ν(C$\equiv$C)	2250–2100	vs
C=C stretch of alkenes	ν(C=C)	1750–1450	vs–m
C–C stretch of aliphatic chains and cycloalkanes	ν(C–C)	1150–950	s–m
CC stretch of aromates [(substituted) benzene molecules]	ν(CC)	1600, 1580, 1500, 1450, 1000	s–m m–w s
C=O stretch	ν(C=O)	1870–1650	s–w
Antisymmetric C–O–C stretch	ν_{asym}(COC)	1150–1060	w
Symmetrical C–O–C stretch	ν_{sym}(COC)	970–800	s–m
CH$_2$ bending vibrations, antisymmetric CH$_3$ bend	δ(CH$_2$), δ_{asym}(CH$_3$)	1470–1400	m

(*continued overleaf*)

Table 1.1   (*continued*).

Vibration		Wavenumbers (cm^{-1})	Raman-intensity[1]
Symmetric CH_3 bend	$\delta_{sym}(CH_3)$	1380	m−w
CH_2 in phase twist	$\delta(CH_2)$	1305−1295	m
Symmetric bending of aliphatic chains (C_n with n=3...12) (=chain expansion)	$\delta_{sym}(CC)$	425−150	s−m
Amide I band (CONH stretch)[2]	Amide I	1670−1630	m−w
Amide III band (C−N stretch, coupled with opening of the CNH angle)[2]	Amide III	1350−1250	m−w

[1] s strong, m medium, w weak, v very (see Intermezzo 1.3).
[2] Vibrations of the bonds in an amide function can not be considered independently. The amide I and III bands are the most important Raman active vibrations of this functional group. Group frequencies of this functional group can be defined more precisely when primary, secondary or tertiary amides are considered separately.

The force constant of the bond is also influenced by the hybridisation state of the considered atoms. In Table 1.1 this can be seen when comparing the band positions of the $\nu(C-H)$ stretching vibration for alkanes, alkenes and alkynes. In the first case, the carbon atom is sp^3 hybridised, whereas in the other cases its hybridisation is sp^2 and sp, respectively. We can see that, in this series, the relative percentage s-character of the hybridised carbon atom rises. The higher the percentage s-character of the carbon atom, the higher the force constant κ.

1.4 Raman Intensity

The overall intensity of a Raman band is the result of a whole series of influences. Basically, they can be divided into two groups:

$$I = \alpha \cdot \beta \qquad (1.25)^2$$

where α refers to parameters related to the sample and β are parameters that are instrument-related. Jawhari *et al.* (T. Jawahari, P.J. Hendra, H.A. Wills and M. Judkins, *Spect. Chim. Acta A* **46**, 161–70 (1990)) express this last group of parameters as:

$$\beta = B \cdot V \cdot F_{instr} \qquad (1.26)$$

where B is related to the intensity of the source of radiation, V is the analysed volume and F_{instr} is a group of instrument-dependent factors, like detector efficiency or measurement geometry (see also Chapter 2). In this paragraph, Raman intensity is considered from a fundamental point of view, and instrumental factors are not taken into account here.

In general, the intensity of scattered light equals the change in electromagnetic flux $d\Phi$, considered over a specific solid angle Ω:

$$I = d\Phi/d\Omega \qquad (1.27)$$

Intensity has as units $W \cdot steradian^{-1}$. The solid angle Ω describes a circle with area A, on a distance r of the source of the scattered radiation, resulting in:

$$dA = r^2 \cdot d\Omega \qquad (1.28)$$

For a specific considered area A (e.g. the size of the collection optics), the measured intensity is inversely proportional to the square of the distance to the source.

---

[2]Here, the symbol α represents a group of parameters, not polarisability.

The intensity of the light that is scattered in a certain direction by a dipole with a specific orientation in space is given by:

$$I = \frac{d\Phi}{d\Omega} = \frac{\pi^2 \cdot c \cdot \bar{\nu}^4 \cdot p_0^2 \cdot \sin^2\theta}{2 \cdot \varepsilon_0} \tag{1.29}$$

Here, $\bar{\nu} = \frac{1}{\lambda}$ is the *absolute* wavenumber[3] of the electromagnetic radiation, p_0 is the amplitude of the induced oscillating dipole moment, c is the speed of light and ε_0 (= $8.854187817 \cdot 10^{-12}$ C^2 · N^{-1} · m^{-2}) is the permittivity of vacuum. θ is the angle between the direction of the dipole moment and the direction in which the measurement is done. In reality, intensity is not measured in exactly one direction, but should be integrated over a solid angle Ω (centred around θ) and this expression should be integrated over this solid angle.

$$\Delta\Phi = \frac{\pi^2 \cdot c \cdot \bar{\nu}^4 \cdot p_0^2}{2 \cdot \varepsilon_0} \int_{\theta-\Delta\theta}^{\theta+\Delta\theta} \int_{\varphi-\Delta\varphi}^{\varphi+\Delta\varphi} \sin^3\theta \, d\theta \, d\varphi \tag{1.30}$$

If $\bar{\nu}$ is expressed in cm^{-1} in Equations (1.29) and (1.30), p_0 in C·m, Φ in W and I in W · sr^{-1}, then we obtain the following expression, if all constants are grouped:

$$I = 1{,}6709 \cdot 10^{28} \cdot \bar{\nu}^4 \cdot \sin^2\theta \cdot p_0^2 \tag{1.29'}$$

$$\Delta\Phi = 1{,}6709 \cdot 10^{28} \cdot \bar{\nu}^4 \cdot p_0^2 \int_{\theta-\Delta\theta}^{\theta+\Delta\theta} \int_{\varphi-\Delta\varphi}^{\varphi+\Delta\varphi} \sin^3\theta \, d\theta \, d\phi \tag{1.30'}$$

In Equations (1.30) and (1.30'), p_0 is independent of the orientation. It can be seen that the measured intensity is proportional to $\bar{\nu}^4$, and as a consequence, the Raman signal is seriously stronger, the shorter the laser wavelength that is used – certainly a reason

---

[3] In Equation (1.29) the absolute wave number is used $\bar{\nu} = \frac{1}{\lambda} = \frac{\nu}{c}$, opposite to the Raman wave number $\bar{\nu}_v = \frac{\nu_v}{c}$. The relation between both is: $\bar{\nu} = \bar{\nu}_0 \pm \bar{\nu}_v = \frac{\nu_0 \pm \nu_v}{c}$, with ν_0 laser frequency.

why UV-Raman spectroscopy is an interesting technique. From Equations (1.29) to (1.30′) it can be seen that the measured intensity is proportional to the square of the amplitude of the polarisability and we see as well a strong dependence on the orientation ($\sin^2\theta$) of the measured Raman effect. Apart from this, the polarisability α is also dependent on the direction, as this is basically a tensor (i.e. an operator that changes one vector in another). The direction of the induced dipole is not always the same as those of the electrical field vector. In a Cartesian coordinate system, Equation (1.2) can be written as:

$$
\begin{pmatrix} \mathbf{p}_x \\ \mathbf{p}_y \\ \mathbf{p}_z \end{pmatrix} = \begin{pmatrix} \alpha_{xx} & \alpha_{xy} & \alpha_{xz} \\ \alpha_{yx} & \alpha_{yy} & \alpha_{yz} \\ \alpha_{zx} & \alpha_{zy} & \alpha_{zz} \end{pmatrix} \cdot \begin{pmatrix} \mathbf{E}_x \\ \mathbf{E}_y \\ \mathbf{E}_z \end{pmatrix} \tag{1.31}
$$

with the vectors $\mathbf{p}$ and $\mathbf{E}$ expressed as their respective components $\mathbf{p}_x, \mathbf{p}_y, \mathbf{p}_z$ en $\mathbf{E}_x, \mathbf{E}_y, \mathbf{E}_z$. The polarisability tensor α is expressed in a matrix. Usually, the polarisability tensor is symmetrical ($\alpha_{ij} = \alpha_{ji}$), which requires us to consider 6 components. However, two properties are independent of the axes chosen: the average polarisability $\overline{\alpha}$ and the anisotropy factor γ[4]. These variables are defined as:

$$
\overline{\alpha} = \frac{1}{3} \cdot (\alpha_{xx} + \alpha_{yy} + \alpha_{zz}) \tag{1.32}
$$

$$
\gamma^2 = \frac{1}{2} \cdot [(\alpha_{xx} - \alpha_{yy})^2 + (\alpha_{yy} - \alpha_{zz})^2 + (\alpha_{zz} - \alpha_{xx})^2
$$
$$
+ 6 \cdot (\alpha_{xy}^2 + \alpha_{xz}^2 + \alpha_{yz}^2)] \tag{1.33}
$$

The polarisability tensor can be split into two components: an isotropic tensor α_{iso} and an anisotropic tensor α_{aniso}.

α_{iso} can be expressed as a diagonal matrix, with the diagonal elements equalling $\overline{\alpha}$. In α_{aniso} the diagonal elements equals $\alpha_{ii} - \overline{\alpha}$,

---

[4] Don't confuse with the hyper polarisability, for which the same symbol is used (Equation (1.1)).

while the other elements are equal to the respective components α_{ij}. As a consequence, the polarisability can be expressed as a sum:

$$\alpha = \alpha_{\text{iso}} + \alpha_{\text{aniso}} \qquad (1.34)$$

Totally isotropic systems have an anisotropy γ of 0 and the anisotropic polarisability tensor $\alpha_{\text{aniso}} = 0$, since the mixed terms α_{ij} are equal to 0.

Let's consider a molecule with a fixed position and orientation in space. We choose our axis system so that the molecule is in the origin of the axis system and the incident beam is along the x-axis (Figure 1.3). In Raman spectroscopy, two types of measurement geometry are often encountered:

- the traditional 90° geometry, where the scattered light is measured perpendicular to the incident beam;
- the back-scattering geometry, where the light is measured that is back-scattered in the direction of the laser.

The electrical field vector **E** of the incident light can be decomposed in two vectors, perpendicular to each other: $\mathbf{E}_y$ and $\mathbf{E}_z$; since the field vector is perpendicular to the direction of propagation, the third component $\mathbf{E}_x = 0$. For unpolarised light, in a certain period of time, the average amplitudes for both components have equal sizes.

For Rayleigh scattering, in the 90° geometry, we can state that:

$$\mathbf{p}_y = \alpha_{yx} \cdot \mathbf{E}_x + \alpha_{yz} \cdot \mathbf{E}_z \qquad (1.35)$$

$$\mathbf{p}_z = \alpha_{zx} \cdot \mathbf{E}_x + \alpha_{zz} \cdot \mathbf{E}_z \qquad (1.36)$$

When considering Raman scattering, the derivative of the polarisability to the normal coordinate has to be considered (i.e. the second term in Equation (1.18)). Similar equations as (1.35) and

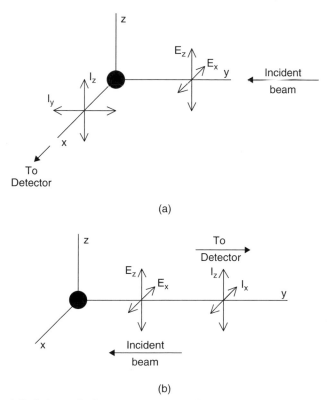

(a)

(b)

Figure 1.3 Schematical representation of two traditional measurement geometries: (a) 90° geometry and (b) back-scattering geometry.

(1.36) can be obtained, with the replacement of the polarisability with its first derivative. In back-scattering geometry, for Rayleigh scattering, we find:

$$\mathbf{p}_x = \alpha_{xx} \cdot \mathbf{E}_x + \alpha_{xz} \cdot \mathbf{E}_z \tag{1.37}$$

$$\mathbf{p}_z = \alpha_{zx} \cdot \mathbf{E}_x + \alpha_{zz} \cdot \mathbf{E}_z \tag{1.36}$$

When considering a molecule with a specific orientation in space (e.g. when describing a monocrystal), the intensity of the scattered radiation as a function of the orientation can be described as: (note that, for unpolarised light: $\mathbf{E}_x = \mathbf{E}_z = \mathbf{E}$):

- in 90° geometry:

$$I_{y\,(\text{Rayleigh})} = \text{Const} \cdot (\alpha_{zy}^2 \cdot \mathbf{E}_z^2 + \alpha_{xy}^2 \cdot \mathbf{E}_x^2)$$
$$= \text{Const} \cdot (\alpha_{zy}^2 + \alpha_{xy}^2) \cdot \mathbf{E}^2 \qquad (1.38)^5$$

$$I_{z\,(\text{Rayleigh})} = \text{Const} \cdot (\alpha_{zz}^2 \cdot \mathbf{E}_z^2 + \alpha_{xz}^2 \cdot \mathbf{E}_x^2)$$
$$= \text{Const} \cdot (\alpha_{zz}^2 + \alpha_{xz}^2) \cdot \mathbf{E}^2 \qquad (1.39)$$

$$I_{y\,(\text{Raman})} = \text{Const} \cdot (\alpha_{zy}'^2 \cdot \mathbf{E}_z^2 + \alpha_{xy}'^2 \cdot \mathbf{E}_x^2)$$
$$= \text{Const} \cdot (\alpha_{zy}'^2 + \alpha_{xy}'^2) \cdot \mathbf{E}^2 \qquad (1.40)$$

$$I_{z\,(\text{Raman})} = \text{Const} \cdot (\alpha_{zz}'^2 \cdot \mathbf{E}_z^2 + \alpha_{xz}'^2 \cdot \mathbf{E}_x^2)$$
$$= \text{Const} \cdot (\alpha_{zz}'^2 + \alpha_{xz}'^2) \cdot \mathbf{E}^2 \qquad (1.41)$$

- in back-scattering geometry:

$$I_{z\,(\text{Rayleigh})} = \text{Const} \cdot (\alpha_{zz}^2 \cdot \mathbf{E}_z^2 + \alpha_{xz}^2 \cdot \mathbf{E}_x^2)$$
$$= \text{Const} \cdot (\alpha_{zz}^2 + \alpha_{xz}^2) \cdot \mathbf{E}^2 \qquad (1.39)$$

$$I_{x\,(\text{Rayleigh})} = \text{Const} \cdot (\alpha_{zx}^2 \cdot \mathbf{E}_z^2 + \alpha_{xx}^2 \cdot \mathbf{E}_x^2)$$
$$= \text{Const} \cdot (\alpha_{zx}^2 + \alpha_{xx}^2) \cdot \mathbf{E}^2 \qquad (1.42)$$

$$I_{z\,(\text{Raman})} = \text{Const} \cdot (\alpha_{zz}'^2 \cdot \mathbf{E}_z^2 + \alpha_{xz}'^2 \cdot \mathbf{E}_x^2)$$
$$= \text{Const} \cdot (\alpha_{zz}'^2 + \alpha_{xz}'^2) \cdot \mathbf{E}^2 \qquad (1.41)$$

[5]In these equations, a series of parameters are grouped as '*Const*'. This term is different for Rayleigh or Raman scattering. It is, amongst others, function of the used laser wavelength. See also Equation (1.29).

$$I_{x\,(\text{Raman})} = \text{Const} \cdot ({\alpha'_{zx}}^2 \cdot \mathbf{E}_z^2 + {\alpha'_{xx}}^2 \cdot \mathbf{E}_x^2)$$

$$= \text{Const} \cdot ({\alpha'_{zx}}^2 + {\alpha'_{xx}}^2) \cdot \mathbf{E}^2 \qquad (1.43)$$

The total intensity can be considered as the sum of the corresponding expressions. For $90°$ geometry $(I_{\text{tot},x})$ and for back-scattering geometry $(I_{\text{tot},y})$ the following expressions can be obtained, respectively:

$$I_{\text{tot},x\,(\text{Rayleigh})} = \text{Const} \cdot (\alpha_{zy}^2 + \alpha_{xy}^2 + \alpha_{xz}^2 + \alpha_{zz}^2) \cdot \mathbf{E}^2 \quad (1.44)$$

$$I_{\text{tot},x\,(\text{Raman})} = \text{Const} \cdot ({\alpha'_{zy}}^2 + {\alpha'_{xy}}^2 + {\alpha'_{xz}}^2 + {\alpha'_{zz}}^2) \cdot \mathbf{E}^2$$
$$(1.45)$$

$$I_{\text{tot},y\,(\text{Rayleigh})} = \text{Const} \cdot (\alpha_{xx}^2 + 2 \cdot \alpha_{xz}^2 + \alpha_{zz}^2) \cdot \mathbf{E}^2 \qquad (1.46)$$

$$I_{\text{tot},y\,(\text{Raman})} = \text{Const} \cdot ({\alpha'_{xx}}^2 + 2 \cdot {\alpha'_{xz}}^2 + {\alpha'_{zz}}^2) \cdot \mathbf{E}^2$$
$$(1.47)$$

When considering a molecule with an isotropic polarisability tensor, the terms α_{jk} disappear and the retrieved expression is further simplified.

Discussion Topic 1.2

What happens if we work with linearly polarised light?

Answer:

Depending on the orientation of the plane of polarisation, in relation to the orientation of the molecule, different results can be obtained, considering Equations (1.38)–(1.43). In these cases, $\mathbf{E}_x$ no longer equals $\mathbf{E}_z$. By changing the plane of polarisation, it is possible to make either $\mathbf{E}_x$ or $\mathbf{E}_z$ disappear.

In practice, it is impossible to record a Raman spectrum of a single molecule. The total signal intensity equals the sum of the contributions of all molecules in the analysed volume V. When considering a group of randomly oriented molecules, the intensity has to be averaged over all possible orientations of the system, in relation to the considered electrical field. Typical examples of such systems with randomly oriented molecules are gasses, ideal fluids, finely dispersed powders (without preferential orientation) and amorphous materials. For these products, we need to consider the average squares of the polarisability. For a symmetrical polarisability tensor – thinking on Equations (1.32) and (1.33) – we can consider:

$$\overline{\alpha_{xx}^2} = \overline{\alpha_{yy}^2} = \overline{\alpha_{zz}^2} = \frac{45 \cdot \overline{\alpha}^2 + 4 \cdot \gamma^2}{45} \qquad (1.48)^6$$

$$\overline{\alpha_{yx}^2} = \overline{\alpha_{yz}^2} = \overline{\alpha_{zx}^2} = \frac{\gamma^2}{15} \qquad (1.49)$$

$$\overline{\alpha_{xx}\alpha_{yy}} = \overline{\alpha_{yy}\alpha_{zz}} = \overline{\alpha_{zz}\alpha_{xx}} = \frac{45 \cdot \overline{\alpha}^2 - 2 \cdot \gamma^2}{45} \qquad (1.50)$$

Other product terms equal 0. Now it is possible, by using an analogous approach as for a specifically oriented molecule, to define the Rayleigh- and Raman-intensities for randomly oriented molecules. Indeed, by using Equations (1.44) to (1.47) we obtain:

- for the 90° geometry:

$$I_{y\,(\text{Rayleigh})} = \text{Const} \cdot (\overline{\alpha_{zy}^2} \cdot \mathbf{E}_z^2 + \overline{\alpha_{xy}^2} \cdot \mathbf{E}_x^2)$$

$$= \text{Const} \cdot \frac{2 \cdot \gamma^2}{15} \cdot \mathbf{E}^2 \qquad (1.51)$$

$$I_{z\,(\text{Rayleigh})} = \text{Const} \cdot (\overline{\alpha_{zz}^2} \cdot \mathbf{E}_z^2 + \overline{\alpha_{xz}^2} \cdot \mathbf{E}_x^2)$$

$$= \text{Const} \cdot \frac{45 \cdot \overline{\alpha}^2 + 7 \cdot \gamma^2}{45} \cdot \mathbf{E}^2 \qquad (1.52)$$

[6]Note: the square of the average polarisability $\overline{\alpha}^2 = \frac{1}{9} \cdot (\alpha_{xx} + \alpha_{yy} + \alpha_{zz})^2$ is different from the average of the square of one of the components of the polarisability tensor $\overline{\alpha_{ij}^2}$.

$$I_{y\,(\text{Raman})} = \text{Const} \cdot (\overline{\alpha_{zy}'^{\,2}} \cdot \mathbf{E}_z^2 + \overline{\alpha_{xy}'^{\,2}} \cdot \mathbf{E}_x^2)$$

$$= \text{Const} \cdot \frac{2 \cdot \gamma'^2}{15} \cdot \mathbf{E}^2 \tag{1.53}$$

$$I_{z\,(\text{Raman})} = \text{Const} \cdot (\overline{\alpha_{zz}'^{\,2}} \cdot \mathbf{E}_z^2 + \overline{\alpha_{xz}'^{\,2}} \cdot \mathbf{E}_x^2)$$

$$= \text{Const} \cdot \frac{45 \cdot \overline{\alpha}'^2 + 7 \cdot \gamma'^2}{45} \cdot \mathbf{E}^2 \tag{1.54}$$

- for the back-scattering geometry:

$$I_{z\,(\text{Rayleigh})} = \text{Const} \cdot (\overline{\alpha_{zz}^2} \cdot \mathbf{E}_z^2 + \overline{\alpha_{xz}^2} \cdot \mathbf{E}_x^2)$$

$$= \text{Const} \cdot \frac{45 \cdot \overline{\alpha}^2 + 7 \cdot \gamma^2}{45} \cdot \mathbf{E}^2 \tag{1.52}$$

$$I_{x\,(\text{Rayleigh})} = \text{Const} \cdot (\overline{\alpha_{zx}^2} \cdot \mathbf{E}_z^2 + \overline{\alpha_{xx}^2} \cdot \mathbf{E}_x^2)$$

$$= \text{Const} \cdot \frac{45 \cdot \overline{\alpha}^2 + 7 \cdot \gamma^2}{45} \cdot \mathbf{E}^2 \tag{1.55}$$

$$I_{z\,(\text{Raman})} = \text{Const} \cdot (\overline{\alpha_{zz}'^{\,2}} \cdot \mathbf{E}_z^2 + \overline{\alpha_{xz}'^{\,2}} \cdot \mathbf{E}_x^2)$$

$$= \text{Const} \cdot \frac{45 \cdot \overline{\alpha}'^2 + 7 \cdot \gamma'^2}{45} \cdot \mathbf{E}^2 \tag{1.54}$$

$$I_{x\,(\text{Raman})} = \text{Const} \cdot (\overline{\alpha_{zx}'^{\,2}} \cdot \mathbf{E}_z^2 + \overline{\alpha_{xx}'^{\,2}} \cdot \mathbf{E}_x^2)$$

$$= \text{Const} \cdot \frac{45 \cdot \overline{\alpha}'^2 + 7 \cdot \gamma'^2}{45} \cdot \mathbf{E}^2 \tag{1.56}$$

The degree of depolarisation ρ_n[7] is defined as the ratio I_y/I_z and in 90° geometry, it can be measured by using a polariser. From

---

[7] ρ_n: The index n shows that we consider the depolarisation ratio for unpolarised light. It is also possible to consider the degrees of depolarisation $\rho_\parallel$ and $\rho_\perp$ when using planary polarised light, polarised in the *xy* or *yz* plane, respectively (for the geometry: see Figure 1.3).

Equations (1.53) to (1.54) it results that the numerical value of the degree of depolarisation equals:

$$\rho_n = \frac{I_y}{I_z} = \frac{6 \cdot \gamma^2}{45 \cdot \overline{\alpha}^2 + 7 \cdot \gamma^2} = \frac{6 \cdot \gamma'^2}{45 \cdot \overline{\alpha}'^2 + 7 \cdot \gamma'^2} \qquad (1.57)$$

For totally isotropic systems $\rho_n = 0$, while for totally anisotropic systems $\rho_n = 6/7$. The total intensity of the scattered light is given by the sum of its components. For $90°$ geometry and for back-scattering geometry, this is, respectively:

$$I_{\text{tot},x \text{ (Rayleigh)}} = \text{Const} \cdot \frac{45 \cdot \overline{\alpha}^2 + 13 \cdot \gamma^2}{45} \cdot \mathbf{E}^2 \qquad (1.58)$$

$$I_{\text{tot},x \text{ (Raman)}} = \text{Const} \cdot \frac{45 \cdot \overline{\alpha}'^2 + 13 \cdot \gamma'^2}{45} \cdot \mathbf{E}^2 \qquad (1.59)$$

$$I_{\text{tot},y \text{ (Rayleigh)}} = \text{Const} \cdot \frac{90 \cdot \overline{\alpha}^2 + 14 \cdot \gamma^2}{45} \cdot \mathbf{E}^2 \qquad (1.60)$$

$$I_{\text{tot},y \text{ (Raman)}} = \text{Const} \cdot \frac{90 \cdot \overline{\alpha}'^2 + 14 \cdot \gamma'^2}{45} \cdot \mathbf{E}^2 \qquad (1.61)$$

In a Raman spectrum, clearly a difference in intensity can be found depending on the different areas in the spectrum. Usually the Rayleigh line is much more intense than the (Stokes or anti-Stokes) Raman bands are. In practice, the Rayleigh line is strongly suppressed with a filter. From Figure 1.1 it can be seen that the intensities in the Stokes region are more intense than the Raman bands in the anti-Stokes region. This can be explained by the Boltzmann distribution, describing the distribution of the molecules over the different energy levels. To understand this, in expression (1.29) we should not simply consider the amplitude of the induced dipole moment $\mathbf{p}_0$ but rather its quantum mechanical equivalent: the dipole transition moment. For the Rayleigh transition, with the same beginning and final state, the quantum mechanical approach results in similar formulas as the classical approach. Stokes and anti-Stokes transitions have a different starting and final situation. Due to the orientation dependence of the dipole transitions, for Rayleigh, Stokes and anti-Stokes transitions, each time a set of three equations can be described, for the following forms, respectively:

Rayleigh:

$$[p_{x0}^{(1)}]_{b \leftarrow b} = \alpha_{xx} \cdot E_{x0} + \alpha_{xy} \cdot E_{y0} + \alpha_{xz} \cdot E_{z0} \qquad (1.62)$$

Stokes:

$$[p_{x0}^{(1)}]_{v_b+1 \leftarrow v_b} = \sqrt{\frac{(v_b + 1) \cdot h}{8 \cdot \pi^2 \cdot c \cdot \overline{v_v}}}$$
$$\cdot (\alpha'_{xx} \cdot E_{x0} + \alpha'_{xy} \cdot E_{y0} + \alpha'_{xz} \cdot E_{z0}) \qquad (1.63)^8$$

Anti-Stokes:

$$[p_{x0}^{(1)}]_{v_b-1 \leftarrow v_b} = \sqrt{\frac{v_b \cdot h}{8 \cdot \pi^2 \cdot c \cdot \overline{v_v}}}$$
$$\cdot (\alpha'_{xx} \cdot E_{x0} + \alpha'_{xy} \cdot E_{y0} + \alpha'_{xz} \cdot E_{z0}) \qquad (1.64)$$

In these equations, the index 0 refers to the amplitude (=time-independent part) of the dipole transition moment. The index b refers to the first situation. The superscript (1) indicates that the approximation of electrical harmonicity is used in these equations. In other words, this means that Equation (1.2) is used instead of Equation (1.1). A consequence of this approximation is that we only consider transitions with a change in vibrational quantumnumber of $\Delta v = \pm 1$ (for Raman transitions).

---

[8]The first term in this equation is the quantum mechanical analogue of Q_{v0}, the amplitude of the normal vibration.

INTERMEZZO 1.3 RAMAN INTENSITY ANNOTATIONS

In literature, for some molecules, a list of Raman band positions is given. Often, the authors also provide an indication of the relative Raman band intensities in the spectrum. Therefore, a list of abbreviations is used, indicating band intensities ranging from very strong (vs) to very weak (vw). Sometimes authors also include indications like br (broad band) or sh (shoulder). For band intensities, there are two systems commonly used – one system with 5 classes, another one with two more intermediate classes. Although no strict rules exist about when this or another band intensity annotation should be given, we provide here an indicative table, to provide the relative band intensity annotations.

Annotation	Relative intensity (relative to most intense band intensity)
Very strong (vs)	100–90%
Strong (s)	90–75%
Medium to strong (m-s)	75–65%
Medium (m)	65–35%
Medium to weak (m-w)	35–25%
Weak (w)	25–10%
Very weak (vw)	10–0%
Very strong (vs)	100–90%
Strong (s)	90–70%
Medium (m)	70–30%
Weak (w)	30–10%
Very weak (vw)	10–0%

Further Reading

P. Vandenabeele, L. Moens, J. Raman, *Spectrosc.* **43** (2012), 1545–50.

An importance difference between classical and quantum mechanical approximation is in the presence of the factors $(v_b + 1)^{1/2}$ and $v_b^{1/2}$ in the equations of the Stokes and anti-Stokes dipole transition moment, respectively. Note that, as in the expression for the intensity of the scattered light (Equation (1.29)) contains the square of the dipole (transition) moment, there is a multiplication with a whole quantumnumber.

At normal measurement temperatures, most molecules are in the vibrational ground state ($v_b = 0$), although some molecules might be in an excited vibrational state ($v_b = 1, 2, \ldots$). The intensity of a certain Stokes transition ($v_b + 1 \leftarrow v_b$) is proportional to $N_{v_b} \cdot (v_b + 1)$, where N_{v_b} equals the number of molecules in the vibrational state v_b. The occupation of a specific energy level is given by the Boltzmann distribution:

$$N_{v_b} = N \cdot \frac{e^{-E_{v_b}/kT}}{\sum_i e^{-E_i/kT}} = N \cdot \frac{e^{-\left(v_b+\frac{1}{2}\right)hc\overline{v_v}/kT}}{\sum_i e^{-\left(v_i+\frac{1}{2}\right)hc\overline{v_v}/kT}} \qquad (1.65)$$

In this equation, N is the total molecule population, k is the Boltzmann constant ($k = 1.380658_{12} \cdot 10^{-23}$ J·K^{-1}) and T is the temperature (in K). Since the harmonic oscillation approximation is considered, the wavenumbers for the different transitions with $\Delta v = 1$ coincide (*i.e.* $1 \leftarrow 0, 2 \leftarrow 1, 3 \leftarrow 2, \ldots$) and the measured intensity of a certain vibration is proportional to $\sum_{v_b}[N_{v_b} \cdot (v_b + 1)]$. If, in this equation N_{v_b} is replaced by Equation (1.65), we obtain:

$$\sum_{v_b}[N_{v_b} \cdot (v_b + 1)] = \frac{N}{1 - e^{-hc\overline{v_v}/kT}} \qquad (1.66)$$

For anti-Stokes transitions, by an analogy, we obtain:

$$\sum_{v_b}(N_{v_b} \cdot v_b) = \frac{N}{e^{hc\overline{v_v}/kT} - 1} \qquad (1.67)$$

These equations can now be implemented in the general intensity equation. Following Equations (1.59) and (1.61), for a group of N molecules, randomly oriented in space the intensity of the Stokes Raman band in 90° and back-scattering geometry, respectively, can be given by:

$$I_{tot,x(Stokes)} = \frac{h \cdot N \cdot (\overline{\nu_0} - \overline{\nu_v})^4}{8 \cdot \varepsilon_0^2 \cdot c \cdot \overline{\nu_v}} \cdot \frac{45 \cdot \overline{\alpha'}^2 + 13 \cdot \gamma'^2}{45}$$

$$\cdot \frac{\mathbf{E}^2}{1 - e^{-h \cdot c \cdot \overline{\nu_v}/kT}} \tag{1.68}$$

$$I_{tot,y(Stokes)} = \frac{h \cdot N \cdot (\overline{\nu_0} - \overline{\nu_v})^4}{8 \cdot \varepsilon_0^2 \cdot c \cdot \overline{\nu_v}} \cdot \frac{90 \cdot \overline{\alpha'}^2 + 14 \cdot \gamma'^2}{45}$$

$$\cdot \frac{\mathbf{E}^2}{1 - e^{-h \cdot c \cdot \overline{\nu_v}/kT}} \tag{1.69}$$

For anti-Stokes transitions, these expressions can be obtained for 90° and back-scattering geometry, respectively:

$$I_{tot,x(anti-Stokes)} = \frac{h \cdot N \cdot (\overline{\nu_0} + \overline{\nu_v})^4}{8 \cdot \varepsilon_0^2 \cdot c \cdot \overline{\nu_v}} \cdot \frac{45 \cdot \overline{\alpha'}^2 + 13 \cdot \gamma'^2}{45}$$

$$\cdot \frac{\mathbf{E}^2}{e^{h \cdot c \cdot \overline{\nu_v}/kT} - 1} \tag{1.70}$$

$$I_{tot,y(anti-Stokes)} = \frac{h \cdot N \cdot (\overline{\nu_0} + \overline{\nu_v})^4}{8 \cdot \varepsilon_0^2 \cdot c \cdot \overline{\nu_v}} \cdot \frac{90 \cdot \overline{\alpha'}^2 + 14 \cdot \gamma'^2}{45}$$

$$\cdot \frac{\mathbf{E}^2}{e^{h \cdot c \cdot \overline{\nu_v}/kT} - 1} \tag{1.71}$$

From these equations, and in agreement with Figure 1.1, it can be seen that the intensity of anti-Stokes radiation is lower than that of Stokes-Radiation. As a consequence, for analytical purposes, Stokes radiation is usually used.

We see in the first instance a linear relationship between the measured intensity and the number of molecules in the sampled volume, which can form the basis for quantitative measurements. In this approach, we didn't take degeneration into account: the fact that two different vibrations may correspond to the same energy level.

We see that in both cases (90° geometry as well as back-scattering geometry) the ratio between the intensity of Stokes and anti-Stokes scattering equals:

$$\frac{I_{Stokes}}{I_{anti-Stokes}} = \frac{(\overline{\nu_0} - \overline{\nu_k})^4}{(\overline{\nu_0} + \overline{\nu_k})^4} \cdot e^{\frac{h \cdot c \cdot \overline{\nu_v}}{k \cdot T}} \tag{1.72}$$

In theory, we see that we can use this expression to use our Raman spectrometer as an expensive thermometer.

Discussion Topic 1.3

How can we measure the temperature of analysis from a Raman spectrum?

Answer:

By using Equation (1.72), it is possible to calculate the temperature T. For a given Raman band, at band the ratio is made of the Stokes intensity at position $\overline{\nu_k}$ and the anti-Stokes intensity at position $-\overline{\nu_k}$. The laser wavelength, expressed in wavenumbers, is also used in this expression. The only nonconstant parameter is the temperature T, which can thus be calculated.

In practice, this approach is hampered by the occurrence of wavelength-dependent sensitivity of the detector, which has to be taken into account when making the intensity ratio. Also, someone has to correct for the occurrence of fluorescence and eventually also for wavelength-dependent self-absorption of the sample. Moreover, when irradiated with laser light, the sample may warm up, resulting in a changing occupation of the energy levels. Especially if Stokes and anti-Stokes radiation are not measured simultaneous, this may seriously influence the estimation of the temperature.

Recording a Raman spectrum can sometimes be hampered by the occurrence of absorption, by the analyte molecule or by the matrix. A typical example of this interference is the influence of water on the spectrum. The three normal vibrations of water are Raman active, but the bands can clearly be distinguished, and unless infrared spectroscopy, little interference is expected. However, water strongly absorbs radiation in the infrared region of the spectrum and when recording a spectrum in an aqueous solution, one should take these absorption effects into account. One the one hand, laser light with wavenumber $\overline{\nu_0}$ can be absorbed, while on the other hand the scattered light with wavenumber $\overline{\nu} = \overline{\nu_0} \pm \overline{\nu_v}$ may be absorbed as well.

In some other cases, the Raman scattered light is much more intense than initially expected. Enhancement of the Raman signal can be caused by different reasons, often related to time-dependent effects. The previous equations are not able to deal with this, since only the amplitude of the electrical field (i.e. the time-independent part) was involved. One of the most important time-dependent effects is the Resonance Enhanced Raman effect. Intuitively, we can understand this from Figure 1.2: the schematical diagram of the different transitions. The (in-)elastic scattering of the light is presented as an absorption-emission mechanism. In this case, the molecule reaches during a short time an excited state $\Delta E = h \cdot \nu_0$ higher than the initial state. This virtual energy level is usually situated in a 'forbidden' zone. However, in the case of resonance enhancement, this energy level coincides with a stable electronic state. As the absorption process in this case can be considered as an 'allowed' transition, the absorption-emission process has a higher probability of occurrence, which means that we can expect a higher Raman intensity than initially thought.

1.5 Raman Bandwidth

The Raman bandwidth is determined by a number of factors. Firstly, there is natural band broadening, which is a direct consequence of the Heisenberg uncertainty principle ($ca.\ 10^{-8}\ cm^{-1}$). Raman spectra of gasses may also undergo Doppler broadening ($ca.\ 10^{-3}\ cm^{-1}$ at 300 K), which is dependent on temperature and pressure. Most spectrometers have insufficient resolution to detect these effects.

However, for most materials – being amorphous or nonideal single chrystals – the local molecular neighbourhood is the most important source of band broadening. From Equation (1.23) we know that the Raman band position in a spectrum is determined by two factors: the reduced mass of the atoms and the force constant of the bond. Chemical environment does not influence the reduced mass; however, it does influence the force constant of the chemical bond. A different chemical environment may cause a shift in the Raman band position. If each of these bonds present in the measured volume of the sample does not have the same chemical environment, there may be a range of slightly different force constants, which consequently may cause band broadening. This clearly explains why Raman bands of amorphous materials are broader than those of crystalline materials.

The second parameter in Equation (1.23) that determines the band position is the mass of the contributing atoms. Differences in atomic mass (i.e. different isotopes) give rise to different Raman band positions. If these different Raman bands are closer than the spectral resolution of the spectrometer, the Raman bands are observed as a single (broader) band.

For Raman spectra of gasses and fluids, another type of Raman band broadening is to be considered: in these materials, also rotation vibration transitions arc possiblc. During a strctching vibration, the bond length varies as a cosinus equation with time (Equation (1.6)). The angular momentum (and thus the rotational energy) of the molecule is dependent on the bond distance. Thus, rotational and vibrational energy of the molecule are coupled; for each vibrational transition, there are different roto-vibrational transitions, allowing the energy to be raised or lowered with a number of rotational quantum steps. Rotational quantum steps are much lower than vibrational transitions, so that each fundamental vibrational transition can be associated with a number of roto-vibrational bands, being slightly lower or higher than the fundamental transition. Again, if the spectral resolution of the spectrometer is insufficient to resolve them, this causes band broadening. Therefore, freezing liquids sometimes allows to obtain sharper Raman bands. However, a different aggregation state may cause sharper bands, but sometimes at different positions in the Raman spectrum, for instance due to the occurrence of hydrogen bonding. The change in molecular environment again changes the force constant of the bond and consequently of the Raman band position.

Apart from these theoretical aspects, instrumental parameters have an influence on the Raman bandwidth. An important characteristic is the spectral resolution of the spectrometer. In dispersive systems, this is mainly determined by the line density of the spectrometer grating, the size of the entrance slit, the pixel size of the CCD detector and the distance between grating and detector (focal length of the spectrometer). For Fourier-transform spectrometers, the spectral resolution is mainly determined by the travel distance of the moving mirror in the interferometer, as this determines the number of fringes per wavenumber in the interferrogram (see Chapter 4, Section 4.4.2). Apart from this, the wavelength stability of the laser is of importance. This can be influenced by the working temperature of the system, and often lasers are equipped with an internal temperature calibration system. Laser side-bands are removed by using a filter system.

1.6 The General Appearance of a Raman Spectrum

QUESTION 1.4

Complete the following paragraph.

Every Raman spectrum (Figure 1.1) has a very intense band at $0\ cm^{-1}$: the line, caused by _____ scattering of electromagnetic radiation. Normally the intensity of this line is suppressed by using appropriate filters.

At positive wavenumbers, _____ Raman bands are observed, while at negative wavenumbers, one can find _____ Raman bands. We know that _____ Raman scattering is more intense than _____ scattering, as a consequence of the _____ distribution.

Stokes and anti-Stokes bands are positioned symmetrically to the Rayleigh line. The observed Raman bands correspond with transitions for which there is a change in _____ of the molecule. By using group theory, it is possible to study whether certain transitions will give rise to a Raman band or not.

The Raman band position is dependent on the _____ of the bond, as well as of the _____ of the constituting atoms. As a consequence, _____ can be defined. Bonds with relatively heavy elements (typically inorganic materials, such as metal oxides) give rise to Raman bands at relatively _____ wavenumbers. Moreover, crystalline materials can undergo lattice vibrations: these can hardly be considered as intramolecular vibrations, but are merely vibrations of larger units relative to

---

each other. The corresponding Raman bands are observed at _____ wavenumbers and are strongly dependent on the local symmetry. Amorphous materials give in general rise to _____ Raman bands, compared to crystalline materials.

1.7 Summary

The theoretical aspects of Raman spectroscopy have been introduced in this chapter. We have discussed the selection rule for Raman spectroscopy, namely that during a Raman active vibration, a change in polarisability occurs. To be able to interpret Raman spectra, it is important to know the different factors that determine the position of a Raman band and we have discussed the concept of group frequencies. In this chapter, also the influence of the laser wavelength on the Raman spectra has been described and we have determined the different factors that influence the Raman bandwidth.

Further Reading

Banwell, C.N., *Fundamentals of Molecular Spectroscopy*, 3rd Edition, McGraw-Hill, London, UK, 1983.
Ferraro, J.R., Nakamoto, K., Brown, C.W., *Introductory Raman Spectroscopy*, 2nd Edition, Academic Press, Amsterdam 2003.
Long, D.A., *Raman Spectroscopy*, McGraw-Hill, Maidenhead, UK, 1977.
Smith, E. and Dent, G., *Modern Raman Spectroscopy: A Practical Approach*, John Wiley & Sons Ltd, Chichester, UK, 2005.

Chapter 2
Interferences and Side-effects

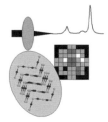

Learning Objectives

- To know the other interactions of electromagnetic radiation with matter, apart from the Raman effect
- To understand the principles of how to avoid possible interferences
- To appreciate the influence of the laser wavelength on the occurrence of fluorescence

If light (electromagnetic radiation) interacts with matter, different phenomena may occur, like scattering, absorption and transmission. The Raman effect is a relatively weak effect, compared to some other phenomena, like fluorescence, which may cause strong interference. Moreover, if laser intensity is too high, the sample may heat (eventually leading to thermal decomposition), or photo decomposition or even ablation may occur. Apart from interferences caused by interaction of the laser light with the sample, background radiation or ambient light may interfere with the measurements. In this section, different phenomena will be discussed.

Practical Raman Spectroscopy: An Introduction, First Edition. Peter Vandenabeele.
© 2013 John Wiley & Sons, Ltd. Published 2013 by John Wiley & Sons, Ltd.

2.1   Absorption

When the wavelength of the laser corresponds with an absorption band in the spectrum of the molecule or its matrix, the laser light is absorbed and the molecules are excited towards an excited state. Depending on the type of radiation on hand, the molecules are excited towards a vibrational excited state (infrared absorption) or towards an electronic excited state (UV–VIS radiation). The absorbed energy is released by radiationless transitions to heat, that is transferred to the environment. It is clear that the absorption strongly interferes with the Raman effect, as the intensity of the scattered radiation is proportional to the number of incident photons that reach the analyte molecule. To avoid this interference, sometimes one can choose to use a laser with another wavelength. A typical example is the absorption of infrared laserlight by water molecules when trying to record a Raman spectrum of an aqueous solution, by using a 1064 nm Nd:YAG laser.

It is clear that, apart from the absorption of the laser beam, one should also consider the absorption of the Raman scattered light. In this case the use of a different laser wavelength is advisable, or in some cases the use of anti-Stokes radiation can be considered.

2.2   Fluorescence

Fluorescence is a serious interference when using Raman spectroscopy. The working principle is shown in Figure 2.1.

The molecule is excited by an incident beam to an excited electronic state. By radiationless transitions, the molecule decays to a lower energy level. From this level, the molecule can further decay to the ground electronic state, while emitting radiation. In principle, the emitted radiation has a lower energy as the incident laser beam.

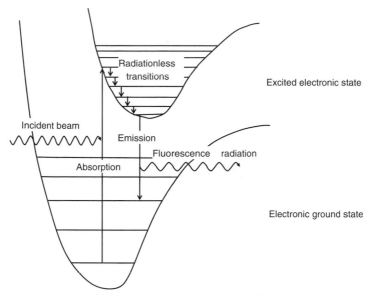

Figure 2.1 Energy diagram of the fluorescence effect.

Interference caused by fluorescence, can be avoided when using a laser with a different wavelength, avoiding the excitation to the excited electronic state. Indeed, when using a laser with a longer wavelength, the energy of the exciting photons is too low to excite the molecule to the excited electronic state. However, as the Raman intensity is proportional to the 4th power of the laser frequency (eq. 1.29), the intensity of the Raman signal also drops when moving towards the infrared laser (not mentioning the wavelength-dependent detector response). Moreover, sometimes it is favourable to excite molecules to the electronic state, because of the occurrence of a resonance effect (see next chapter).

However, if it is not possible to select another laser to avoid fluorescence, sometimes people use the principle of 'photobleaching'. Indeed, fluorescence is always to a certain extent in competition with absorption. On the one hand, the sample will heat, which eventually can lead to thermal decomposition of the fluorophore molecule. On the other hand, photodecomposition may occur, also leading to destruction of the fluorophore group. If the fluorescence is not originating from the analyte molecule, but rather from a compound in the matrix, the sample can be photobleached: someone can try to irradiate the sample for a prolonged time with the laser, before recording the spectrum.

Some instruments use other approaches to avoid fluorescence. One of these approaches is the use of time-dependent measurements. The Raman effect is a relatively fast process, compared to fluorescence or phosphorescence. Indeed, as the Raman effect is an absorption-emission process that goes over a virtual (instable) state, it is relatively fast. Fluorescence is an effect that passes through a series of stable states and the radiationless transitions also need some time. When using a pulsed laser and a gated detector (i.e. a detector with a programmable shutter) one can select the time interval when the detector is open. Thus, it is possible to select the time-frame shortly after the laser pulse when the Raman effect occurs, but when yet no fluorescence is detected.

Another approach to avoid fluorescence is by using mathematical tools to post-process the spectrum. This approach, however, is able to eliminate the fluorescence background from the spectrum, but not the noise associated with this fluorescence (see Chapter 4, Section 4.7). By taking the (first or second) derivative of a Raman spectrum, the fluorescence of the spectrum (but not the noise) can be eliminated. This principle is based on the fact that fluorescence is a broadband feature, in comparison with Raman bands. By making the window for taking the derivatives sufficiently small, broad features are eliminated, while the information on the narrow bands are preserved.

Other procedures have been proposed to avoid fluorescence by recording (at least) two slightly different Raman spectra. This is done either by using two lasers with slightly different wavelengths or by rotating the diffraction grating in the spectrometer. These shifted Raman spectra are then subtracted from each other, resulting in what resembles a derivative of a spectrum. One can fit a first derivative function to this spectrum and by integrating this function the original spectrum is reconstructed. Sometimes, it is claimed that thus not only the fluorescence but also the associated noise is eliminated, but this is merely a smoothing effect caused by the curve fitting.

2.3 Thermal Effects, Photodecomposition and Laser Ablation

The intensity of the Raman signal is proportional to the intensity of the laser beam on the sample. As a consequence, it is advantageous to measure with the laser intensity as high as possible. If the energy input in the system is too high, some destructive effects may occur. Firstly, the sample can heat. This occurs because of absorption of radiation by the sample and if the heat transfer to the surrounding area is too low, the sample heats. Although this effect is observed with all possible lasers, infrared lasers are especially sensitive to this effect.

Secondly, when using lasers with short wavelengths, the energy may correspond with the bonding energy of specific molecular bonds. As a consequence, photo-decomposition may occur.

Laser ablation is a phenomenon where material is removed from the sample. Here, different effects occur simultaneously. It is mainly observed when using pulsed lasers. On the one hand, there is photodecomposition and thermal expansion and shrink (e.g. in the time interval between two pulses). On the other hand, a mechanical effect is present, where the laser radiation exhibits a mechanical force on the sample. When the energy put into the system is larger than the bonding energy and the dissipated energy, ablation occurs and a crater is created.

2.4   Ambient Light and Background Radiation

Since the Raman effect is a weak effect, it is easily overwhelmed by the presence of ambient light or background radiation. We consider here ambient light as radiation present in the room where the measurements take place, where background radiation is radiation caused by the sample (e.g. infrared radiation emitted by a hot sample). It is clear that the presence of ambient light should be avoided and that's why most Raman measurements are performed in a dark room, or at least when shutting the sample chamber from ambient light. Sometimes, it is impossible to perform the measurements in the darkness, and one has to deal with this interference. In this case, also the use of a pulsed laser and a gated detector can be of use, as the detector is only open when the Raman signal is observed. Moreover, a pulsed laser concentrates all energy in a short laser pulse, resulting in a strong Raman signal.

Background radiation cannot be avoided when measuring hot samples, e.g. when recording Raman spectra of a hot melt. These samples emit radiation: black body radiation, which is mainly situated in the infrared region. In these cases, appropriate selection of the laser wavelength (e.g. using blue or UV-lasers) can be useful.

2.5   Summary

In this chapter, different other possible interactions between electromagnetic radiation and matter have been discussed, as these can strongly interfere with Raman spectroscopy. We have also described some common approaches to avoid such interferences. The selection of an appropriate laser wavelength may be crucial to avoid such interferences.

Further Reading

McCreery, R.L., *Raman Spectroscopy for Chemical Analysis*, John Wiley & Sons, Inc., New York, USA, 2000.

Chapter 3
Enhancement of the Raman Signal

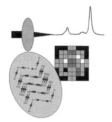

Learning Objectives

- To understand the working principle of the Resonance Raman (RR) effect
- To be able to explain the enhancement principles that form the basis for Surface-Enhanced Raman Spectroscopy (SERS)
- To know some major approaches to prepare substrates for SERS

As already mentioned, the Raman effect is an inherently weak effect. Nevertheless, there are several mechanisms that can be used to enhance the Raman signal. The two most important approaches are resonance enhanced Raman spectroscopy (RR) and surface-enhanced Raman spectroscopy (SERS). These two approaches are explained in more detail in the next paragraphs.

Practical Raman Spectroscopy: An Introduction, First Edition. Peter Vandenabeele.
© 2013 John Wiley & Sons, Ltd. Published 2013 by John Wiley & Sons, Ltd.

3.1 Resonance Raman (RR) Spectroscopy

It can easily be demonstrated that the Raman intensity of a molecule is significantly enhanced when the frequency of the laser corresponds with an allowed transition to an electronic excited state of that molecule. According to the idealised model (see Figure 1.2), this would mean that the virtual state to which the molecule is excited, corresponds with an electronic excited state. Since this (allowed) transition corresponds with a transition to a real state, it is easily understood that this transition is favourable.

The main advantage of this approach is the enhanced sensitivity for specific molecules. On the other hand, this may have the disadvantage that the analyst is focussed on a specific molecule, and that weaker scatterers in the mixture are not easily detected. It is clear that, if someone wants to perform resonance Raman spectroscopy, it is of the utmost importance to select the laser wavelength in function of the molecule on which the research wants to focus.

3.2 Surface-Enhanced Raman Spectroscopy (SERS)

SERS is a technique in which the Raman signal of specific molecules is enhanced by bringing them in contact with certain substrates with a well-defined morphology, usually noble metals (e.g. gold, silver). It is the aim to bring a thin layer of the analyte on the substrate and the Raman signal can in some cases be enhanced with a factor 10^6. This approach is typically used for the analysis of specific molecules, in solution or gas phase (after adsorption to a substrate), which may lead to the application in sensors. Often, molecules with extended conjugated pi-systems, like aromates, are examined. When using SERS, it is of the utmost importance to select the substrate in a way that the SERS effect is maximised. Some specific parameters have to be selected carefully, such as the metal (e.g. Au, Ag), the particle size (when using a colloid) or the roughness. Particularly field enhancement (one of the contributing effects, see below) is high with metals with a high reflectivity (e.g. Au, Ag) and the effect is dependent on the type of material in combination with the laser wavelength on hand. Silver substrates tend to be interesting when used in combination with green or blue lasers, while gold substrates tend to perform better when used with red lasers.

3.2.1 Working Principle of SERS

The SERS enhancement is usually ascribed to two different effects, namely field enhancement and chemical enhancement, where the first factor is thought to be the most important. We give a simplified description of the SERS effect, where we try to explain the main factors in an easy to understand way.

(a) Chemical Enhancement

Chemical enhancement counts for an enhancement effect of up to ca. 100 times. It occurs by interaction of the adsorbed molecule and the metal surface. SERS is a surface-effect: i.e. the molecules close to the surface of the substrate are analysed. As the observed Raman signal is related to the number of molecules that are present in the sampling volume, all interactions between the substrate and the molecules that result in a higher analyte concentration close to the surface enhance the obtained Raman signal. Experiments have been performed with polytetrafluorethylene (PTFE, Teflon) coated glass slides, as these have a different polarity than traditional glass slides and thus may cause some preconcentration of compounds. Moreover, when working in an aqueous solution, due to the surface tension of the water, the droplet does not spread over the whole surface of the slide. Similarly, interactions may take place between metallic surfaces and the analyte molecules. These interactions can be classified as physisorption or chemisorption effects. Physisorption refers to relatively weak interactions (e.g. electrostatic interactions, dipole–dipole effects, Van der Waals interactions, etc.), whereas chemisorption is used to describe the case where a chemical bond is formed between the substrate and the analyte molecule. As a consequence, it may be possible to observe these bonds in the resulting Raman spectrum (different Raman bands are observed compared to the bulk spectra) or bandshifts (compared to the bulk spectrum) may occur due to the interaction.

Another aspect is the orientation effect. As discussed before, for anisotropic molecules, the Raman effect is orientation dependent. Therefore, if the molecules are well aligned, orientation may enhance the Raman effect. Depending on the interaction with the substrate, some molecules can have a preferential orientation, which as such counts for an enhancement effect. For instance, if the substrate interacts with the π-electrons of a benzene molecule, it is likely that the benzene molecule is oriented parallel to the surface of the substrate.

Apart from these preconcentration (chemisorption and/or physisorption) and orientation effects, another chemical enhancement effect may occur, which may be explained by using charge-transfer theory. We will explain this in a rather intuitive way. When describing the resonance Raman effect, we discussed that resonance enhancement occurs when the virtual energy level corresponds with a real state (e.g. an electronic energy level of the molecule). In metals, the orbitals of all the single elements overlap, no longer resulting in relatively narrow energy levels as known for single atoms or even simple molecules. However, we must consider overlapping orbitals, where a certain number is filled with electrons: this is named the conduction band. Because the interaction between the analyte molecule and the energy levels of the molecule may be shifted, as a result of the interaction of the molecule with the metal, the electron cloud of the molecule may be disturbed. Hence electronic energy levels may shift in such a way that the conditions for resonance enhancement are fulfilled. Moreover, it is also possible that during the Raman experiment, there is an excitement towards an energy level in the conduction band of the metal – which is a real state and thus resonance enhancement may occur between the vibrational ground state of the molecule and the conduction band of the metal.

(b) Field Enhancement (see Figure 3.1)

From the theoretical discussion of the Raman effect, we know that the measured Raman signal is proportional to the strength of the oscillating electromagnetic field (i.e. the intensity of the laser beam). As a consequence, if the molecules are irradiated with a more powerful laser, the signal is increased. Field enhancement increases the electromagnetic field strength as experienced by the analyte molecules and occurs if (rough) metal surfaces are exposed to laser light with a specific wavelength. If a small metallic particle is positioned in an oscillating electromagnetic field (like laser light), the electrons on the surface of the particle undergo forces and are moved from one part to another. If the particles are sufficiently small, the electrons oscillate in phase with the electromagnetic radiation and a (dipolar) surface plasmon is formed. These moving electrons generate an electromagnetic field. As the electrons move in phase with the incident laser beam, the generated electromagnetic field is in its turn oscillating with the same frequency as the movement of the electrons, hence the incident laser beam and is also in phase with it. As a consequence, the analyte molecule not only experiences the electromagnetic field of the laser beam, but also interacts with the generated oscillating field. An analogy can be made with the effects as observed from broadcasting antennas: the

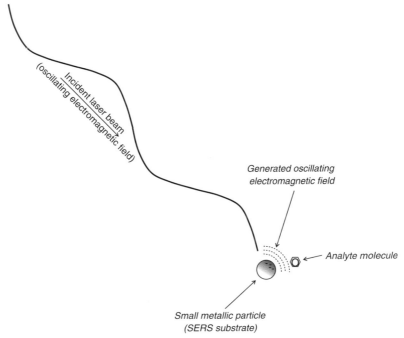

Figure 3.1 Schematic overview of the working principle of field enhancement.

main antenna (laser) emits a signal, that is picked up by an antenna in the field (metal particle). This field antenna (metal particle) emits the signal and acts as an amplifier. The signal is picked up by the antenna on the radio or television set (analyte molecule), where it is transformed into an electrical signal.

From theoretical calculations, based on idealised models (e.g. perfect spherical metal particles, with the same size), it can be understood that the larger the distance between the analyte molecule and the metal particle, the smaller the enhancement is. Indeed, the small metallic particle acts as a spherical source, emitting a spherical electromagnetic field, and the electromagnetic flux diminishes with the third power of the distance to the source. More precisely, the field enhancement is proportional to $(r/d)^3$, where r is the radius of the particle and d is the distance between the centre of the metal particle and the analyte molecule. Therefore, field enhancement is maximal if there is an intense contact between the analyte and the metal surface (r/d equals 1) and drops if the distance becomes larger. However, this reduction of the effect is less severe, the larger the metallic particles are (for a given d, the ratio $(r/d)^3$ is larger if r is larger). However, the field enhancement theory is only valid if the particles are sufficient small, compared to the laser wavelength (typically $r<(\lambda/10)$).

In general, field enhancement strongly depends on the optical properties of the metal. The metal should have a high reflectivity of the wavelength-range of the laser light. The metal and the laser wavelength need to be adapted. In general, Ag is used when performing SERS experiments with blue or green lasers, whilst Au substrates are used with red lasers. The average size of the metal particles, or the roughness of the metal surface, is critical. As mentioned before, particles should be much smaller than the laser wavelength. But, not only is the average size important: also the particle size distribution can seriously influence the results of the experiment. Moreover, their shape is also of importance: two differently shaped particles may have the same average diameter, but may have a totally different curvature of their surfaces.

3.2.2 *Preparation of SERS Substrates*

SERS substrates generally have a limited shelf-life: the substrates degrade over time, for instance as particles aggregate to form larger particles, as metal surfaces might oxidise or as atmospheric compounds might contaminate the surface. Therefore, SERS substrates are typically prepared in the analytical lab before the analysis is performed. Depending on the type of substrate used and the way of preparing it, the substrates can be preserved under an inert atmosphere from a couple of days to a couple of weeks. During the preparation of the SERS substrates, it is of the utmost importance to work with highly pure chemicals and to follow the procedures in a very strict way. If not, the results are hardly predictable: inaccurate procedures might result in differently shaped or sized particles, resulting in lower enhancement factors; impurities may also modify the expected particle shapes, but can also give rise to Raman signals that cannot be attributed to the analyte. In literature, many different procedures are described to produce SERS substrates. A detailed description of these procedures is beyond the scope of this introductory text. However, we will describe here some basic techniques used to produce SERS substrates.

(a) Colloids

A commonly used type of SERS substrate is a colloid of metal particles. As particle size is of the utmost importance to obtain optimal surface enhancement, the particles are kept in suspension and the degree (and speed) of agglomeration is well controlled. During production of the colloid, one typically starts with metal salts that are reduced under controlled circumstances. The quality of the obtained colloid is usually controlled by using UV–VIS absorption spectroscopy – indeed the interaction that causes absorption in the UV–VIS range of the electromagnetic spectrum, is due to transitions between different electronic energy levels and can be related to the phenomena that lead to field enhancement. Typically, the analyte molecules are added to an aqueous suspension of colloidal silver or gold particles or an aliquot of the colloid suspension is spiked to the analyte-bearing solution. Alternatively, the colloid is immobilised on a substrate before the analyte solution is added. Provided the analytical procedure is well performed and the aggregation times are respected, reproducible results can be obtained.

(b) Sputtering

SERS substrates can be prepared by using sputtering. This is a technique where, under an inert atmosphere, a metal target is ablated under bombardment of ions or photons (e.g. laser sputtering). Thus, metal particles are vaporised and precipitate on a surface. When vaporised silver is deposited on, for instance, a glass substrate, the atoms tend to form small aggregates, so-called silver islands. Obviously, this is only the case when the film thickness is small. Although glass substrates are frequently used, other substrates have also been used to provide silver substrates. After this vapour deposition, the samples are cooled and the substrate is exposed to the analyte solution before analysis. As moisture or oxygen in the atmosphere might modify the SERS surface, the substrates are usually preserved under an inert atmosphere or in vacuum.

(c) Electrochemical Production

A third approach to create a roughened metal surface is based on electrochemical processes. In this case, a silver electrode is used. The approach consists of normal electro-analytical procedures, where the polished electrode is brought in an electrochemical cell containing a KCl solution, with a kalomel electrode as counter electrode. The electrode surface is converted into a SERS substrate, by letting the electrode undergo several oxidation-reduction cycles. In the first stage, the silver surface is oxidised towards Ag^+ ions, in the second phase the Ag^+ ions in the solution are reduced at the surface to metallic silver (Ag^0). So, basically, in the first stage of the cycle the metal surface is oxidised and part of the electrode is brought into solution, while in the second half of the cycle, the ions in solution are precipitated at the surface and reduced. It is obvious that the Ag^+ ions that are reduced at the electrode surface will not be well aligned with the electrode surface and that this way a roughened surface is created. After applying several of these oxidation-reduction cycles, the surface is taken out of the cell, rinsed with water and the analyte is brought into contact with the rough metal surface.

(d) Etching

The previous procedure can be considered as an electrochemical way of etching or roughening the metal surface. Similarly, nitric acid can be used to roughen the surface. This approach is far more simple compared to the electrochemical etching, leading to much higher reproducibility. However, the obtained enhancement factors are lower.

3.2.3   SERS Active Molecules

Theoretically, all types of analyte molecules (as far as they are Raman active) can undergo to a certain extent surface enhancement. However, when studying the literature, some molecules are much more sensitive to SERS than others. Thinking on the two types of enhancement, chemical and field enhancement, it is clear that the first type is determined by the type of analyte molecule that is studied, but that the latter type of enhancement is only slightly affected by the analyte molecule. Indeed, field enhancement is affected by the properties of the metal and the laser. Therefore, properties of the molecules under study that interfere with any of the chemical enhancement processes, increase or decrease the surface enhancement. Typically, molecules with loose bound electrons (e.g. conjugated systems or aromates) are more likely to be sensitive to charge-transfer interactions. Also, conjugated systems can also interact with the metal surface, to enhance the signal due to physisorption or even chemisorption. Moreover, due to their shape, some molecules may have a preferential orientation towards the surface, which enhances the signal as well.

3.2.4 Advantages and Disadvantages of SERS

The main advantage of SERS is obviously the enhancement of the Raman signal, and as a consequence the lower limits of detection, compared to normal Raman spectroscopy. The approach has definitely proven its value. However, there are also some disadvantages of this approach. Firstly, the approach requires preparation of a substrate, which is a critical step in obtaining good SERS spectra. If quantitative analysis is required, the request for reproducibility can be critical and requires good laboratory skills. Moreover, it is of the utmost importance to work with (ultra-) pure reagents, as contaminants may either change the properties of the produced surface, either produce their own SERS signal. Secondly, critical in the SERS approach is that the analyte molecule is in close contact with the substrate. For solids, this requirement is a bit more difficult to reach, compared to liquid samples. Sometimes, a droplet of colloid suspension has to be positioned over a solid sample, in order to measure this. Generally, sample preparation is more elaborate than for standard Raman experiments. Thirdly, in a mixture not all molecules yield similar enhancement factors and spectra of certain components may overwhelm the spectrum of other products in the mixture. Finally, the obtained SERS spectra are often hard to compare with standard Raman spectra, which can hamper the spectral interpretation and band assignment.

INTERMEZZO 3.1 THE ANALYSIS OF ORGANIC DYES BY SERS

Dye analysis is a very challenging topic in the analysis of cultural heritage objects. Dyes are – opposite to pigments – colouring agents that are soluble in the binding medium that was used. As a consequence, these colorants are not present as single particles. Dyes are usually of organic nature. In the past, they were typically extracted from plants or insects and are, during their application, fixed on the support. Often, this fixation involves some chemical reaction, like an oxidation, or the use of a mordant: a compound that forms a complex together with the dye molecules, and thus fixes them on the support.

The analysis of these organic dyes is a difficult task. Not only are these colorants typically present in very low concentrations, but they are also usually present in mixtures. Plant extracts contain in general a series of molecules and dyes can be deliberately mixed by the artist. Moreover, during history the dyes may have undergone degradation. When performing Raman spectroscopy, the low concentrations lead to very noisy spectra and fluorescence very often hampers the analysis. As a consequence, it is often not possible to identify the dye in an artwork (e.g. on a tapestry) by using normal Raman spectroscopy.

Therefore, surface-enhanced Raman spectroscopy can be used to enhance the Raman signal. When working on artefacts, sampling is usually very limited, as the object cannot be damaged. As a consequence, it is also not possible to apply a silver colloid on the artwork. Therefore, different approaches have been developed, each trying to minimise the amount of sample that is required. One interesting approach is the use of a bead of a polymer hydrogel, loaded with a mixture of water, organic solvent and complexing agent. This bead is brought into contact with the dye on the art object and after removal of the gel bead, a drop of silver colloid is added to the gel, before analysis under the Raman microscope.

In literature, a whole series of procedures can be found. Examples of dyes that were successfully identified include, amongst others, madder (alizarin), turmeric (curcumin), cochineal (carminic acid), etc. A broad range of artists' materials have been studied, including tapestries, water colour paintings, oil paintings, etc.

Further Reading

Brosseau, C.L., Casadio, F., Van Duyne, R.P., *J. Raman Spectrosc.* 42 (2011), 1305–10.
Brosseau, C.L., Rayner, K.S., Casadio, F., Grzywacz, C.M., Van Duyne, R.P., *Anal. Chem.* 17/81 (2009), 7443–7.
Leona, M., Stenger, J., Ferloni, E., *J. Raman Spectrosc.* 37 (2006), 981–92.
Leona, M., Decuzzi, P., Kubic, T.A., Gates, G., Lombardi, J.R., *Anal. Chem.* 83 (2011), 3990–3.
Pozzi, F., Lombardi, J.R., S. Bruni, M. Leona, *Anal. Chem.* 84 (2012), 3751–7.
Whitney, A.V., Van Duyne, R.P., Casadio, F., *J. Raman Spectrosc.* 37 (2006), 993–1002.

3.3 Summary

In this chapter, two important techniques have been described to enhance the Raman signal. The first approach, resonance enhancement, describes how an appropriate selection of the laser wavelength can cause serious enhancement of the Raman signal. In a second approach, surface-enhanced Raman spectroscopy, the analyte can be brought in close contact with a metal substrate, which can yield to chemical and field enhancement of the Raman signal.

Further Reading

Laserna, J.J (Ed.), *Modern Techniques in Raman Spectroscopy*, John Wiley & Sons Ltd, Chichester, UK, 1996.

McCreery, R.L., *Raman Spectroscopy for Chemical Analysis*, John Wiley & Sons, Inc., New York, USA, 2000.

Smith, E. and Dent, G., *Modern Raman Spectroscopy: A Practical Approach*, John Wiley & Sons Ltd, Chichester, UK, 2005.

Chapter 4

Raman Instrumentation

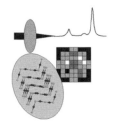

Learning Objectives

- To know the main components of Raman instrumentation
- To appreciate the working principle of a laser
- To know the working principle of a CCD-detector
- To be able to name the different types of filters in Raman spectroscopy
- To understand the working principles of the dispersion systems in dispersive and Fourier-transform (FT-) Raman spectrometers
- To explain the working principle of a fibre-optics probehead and the meaning of its filters
- To understand the origin of noise in Raman spectra

In this part, a summary is given of the most important components of the Raman spectrometer. Although many kinds of Raman spectrometers exist, each with their own advantages and disadvantages, a general diagram can be presented (Figure 4.1).

Practical Raman Spectroscopy: An Introduction, First Edition. Peter Vandenabeele.
© 2013 John Wiley & Sons, Ltd. Published 2013 by John Wiley & Sons, Ltd.

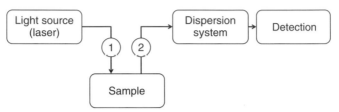

Figure 4.1 General diagram of a (Raman) spectrometer. 1 optics to focus the laser beam on the sample; 2 collection optics for the scattered radiation.

Of course, this diagram is too general to start a detailed study of the different aspects of Raman spectrometers. Components have to be selected according to the tasks and options that are chosen. For example, if a back-scattering geometry is selected, the optics used to focus the laser beam on the sample will be the same as the optics used to transfer the Raman back-scattered light to the spectrometer, where dispersion and detection happen. There are two important types of spectrometers on the market (Figure 4.2), that is, the dispersive spectrometers which use a grating for the separation of the light in its components, and Fourier-transform (FT-) Raman spectrometers, which use a Michelson interferometer. This will be discussed in more detail further in this chapter.

But first, we will discuss the different components of a Raman spectrometer in more detail.

4.1   Lasers

A laser (<u>l</u>ight <u>a</u>mplification through <u>s</u>timulated <u>e</u>mission of <u>r</u>adiation) is an intense source of monochromatic light which can be used for Raman spectroscopy. There are some important characteristics of a laser which make it extremely suitable for this purpose:

- Lasers are intense
- Lasers are monochromatic
- Lasers have a small divergence
- Lasers are polarised
- Lasers are coherent light sources

  In addition, to be suitable for Raman spectroscopy, lasers have to show a high intensity and frequency stability.

  In the following paragraphs we will study the different characteristics of lasers and their working principle. But first, we will discuss different ways to classify them.

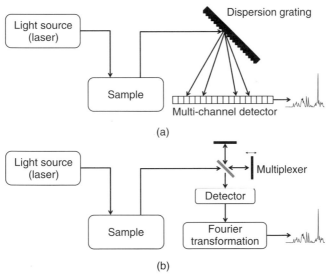

Figure 4.2 General diagram of a dispersive Raman spectrometer (a) and a Fourier-transform (FT-) Raman spectrometer (b).

4.1.1 Classification According to Safety Category

Lasers produce an intense, monochromatic light beam and when they hit a material, they can interact in different ways (see Chapter 2). Thermal, photochemical and mechanical effects (e.g. ablation) can be harmful for the human health, especially for the eyes and skin.

The possible eye damage, besides the intensity, also depends on the laser wavelength. Infrared radiation is mainly absorbed by the moisture which is found on the eye lens, while the UV light is absorbed by the lens itself. Therefore these lasers will mainly damage these parts of the eye. Visible light is focused by our eye lens on our retina. The eye is particular sensitive in this visible range of the electromagnetic spectrum as this type of laser beam is focused in the eye, and therefore an amplification of 100 000 times occurs. As a result of this a visible laser light will damage the retina. Most of the time the victim will not immediately notice the damage; pain and limited vision will manifest itself after some time. It can take several months before the original vision is restored (almost completely). The focus effects do not apply to the skin; therefore the damage to the skin is usually limited. For certain lasers, however, burning may occur which can result in a pigmented spot.

Depending on the effects on health, lasers can be classified into different categories:

- *Category 1 lasers:* are considered safe.
- *Category 2 lasers:* visible lasers, with limited intensity. Briefly looking into the laser beam for less than 0.25 s is not considered harmful. However, intentionally looking into the laser beam is.
- *Category 3 lasers:* can emit any wavelength, its diffuse reflection on a surface does not cause any harm, except after looking for a long time from a short distance. No increased risk for burning or harm to the skin is expected.

Any *continuous wave* laser (see below) with a capacity lower than 0.5 W is category 3 laser or lower. It is harmful to look into a category 3 laser, and laser precautions need to be taken to make this impossible.

- *Category 4 lasers:* a category 4 laser has a higher power than category 3 lasers. These lasers can cause skin and eye damage, even after diffuse reflection.

Working in a laser laboratory requires a few safety measures. It is good practice to limit free running laser beams and to block them with nonreflecting screens. During outlining of optics it is necessary to wear safety goggles. These are equipped with specific filters, which block the light of this specific wavelength and are transparent for other wavelengths. Special attention should go to reflecting surfaces, when brought in the laser beam. A typical hazard is created when the laser beam hits components of the optical bank (e.g. holders of lenses or mirrors, the back of a mirror, ...) whereby the reflecting beam is sent uncontrolled through the room. It is obligatory to have a responsible person for laser safety, from laser category 3b or higher. In addition, regular safety drills need to be organised.

4.1.2 The Operating Principle of the Laser

Lasers are very intense, monochromatic light sources. Various types of lasers are available on the market, and depending on their specific optical characteristics, the desired power and wavelength, one or other type is selected.

The laser's performance is based on the amplification of the signal through stimulated emission. The result of the stimulated emission is that all atoms[1] in the laser together emit electromagnetic

---

[1] Note that here we use the word 'atom', where this could be another type of particle. Dependent on the type of laser these particles can be atoms, molecules, ions, etc.

radiation whereby an intense, coherent laser beam is formed. When an atom is found in an electronically excited state, and when it is irradiated with photons of which the energy equals the difference in energy between the ground state and the excited state, this atom will emit two photons with the same energy. Because of this stimulation, both photons have the same frequency (they have the same energy) and are in phase (they are coherent). Therefore, the total intensity of the emitted wave is intense as no interference between the two emitted photons happens. To maximise this effect it is useful to have many atoms/molecules in the excited state. For this reason in lasers a system of population inversion is created. Indeed, if you use a system of spontaneous emission of radiation, you could theoretically expect that the emitted photon provokes a cascade of stimulated emission. In equilibrium conditions, this does not happen, because the odds are very low that the first emitted photon hits a photon in an excited state. Therefore, population inversion is required.

Usually a lower energy level is more occupied than a higher situated level (cf. Boltzmann Distribution). A system with population inversion can be formed in two ways: either by increasing the number of atoms in the excited state, or by decreasing the number of atoms in ground state. According to the Boltzmann Distribution, the addition of thermal energy to the system will increase the average energy of the system, but will not cause a population inversion; one works under thermodynamic equilibrium conditions. The population inversion can be realised by pumping light or electrical energy to the system.

A very common approach is for the atom to be excited to a higher situated energy level, after which it drops down to the energy level of the laser. Intrinsically the population inversion is a kinetic effect, whereby the excitation to the highest level and the decay to the laser level are intrinsically quick processes (Figure 4.3). The highest laser level is a metastable condition whereby the population is highly populated at this energy level and a population inversion

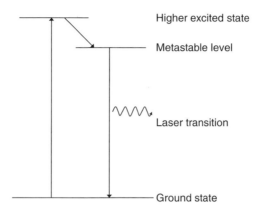

Figure 4.3 3-level energy diagram for laser excitation.

occurs. Typically, the retention time in this metastable condition is 1000 times longer than in the short-lived excited condition. Indirect excitation can be used to excite atoms in an environmental gas mixture, which afterwards transfer their energy to the molecules which cause the laser light.

In practice 4-level laser excitation is often used (Figure 4.4). In this case the population inversion is realised between the metastable level and an excited condition. Thanks to the Botzmann distribution the lowest laser level is less populated than the ground state, whereby the population inversion occurs more easily. Furthermore the atoms will fall back from the lowest laser level to the ground state, which enables continuous operation.

Apart from the population inversion, there should also be a cavity to amplify the light in that way. Light which was produced by stimulated emission in a laser medium usually has one single wavelength, but needs to be taken out of the medium by an amplifying mechanism. This is done by the resonance cavity which reflects the light back in the laser and amplifies the light intensity by various interactions (Figure 4.5). In this way, after stimulated emission, two

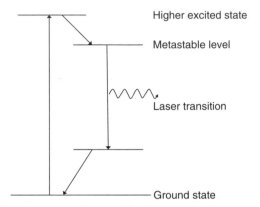

Figure 4.4 4-level energy diagram for laser excitation.

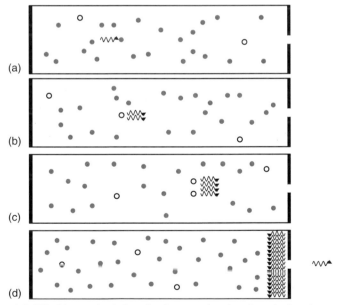

Figure 4.5 Working principle of a laser: stimulated emission in a cavity.

photons are formed with the same frequency, which in turn can de-excite other excited atoms and therefore generate new photons, each with the same energy and phase angle. The number of photons that are generated are proportional to the path length of the light in the laser medium.

INTERMEZZO 4.1 THE HELIUM-NEON LASER

The Helium-Neon laser, is one of the most important gas lasers. Its working principle can be understood by looking at the energy diagram in Figure Intermezzo 4.1).

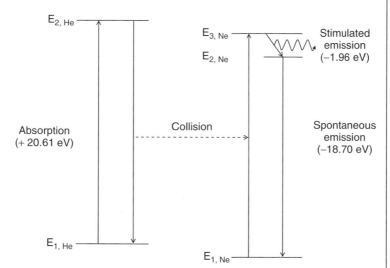

Figure Intermezzo 4.1 The Helium-Neon laser.

By electrical discharge, Helium atoms can be brought to the excited energy level $E_{2,He}$, which is 20.61 eV above the ground state $E_{1,He}$. Neon atoms can be excited from their ground state ($E_{1,Ne}$) to the $E_{3,Ne}$ excited state by collision with the excited He

---

atoms, as the energy that is required for this excitation almost equals the energy difference between the $E_{2,He}$ and $E_{1,He}$ states. The small excess of energy that is required can be gained from the kinetic (translation) energy of the atoms. One can state that Helium 'pumps' Ne atoms towards the $E_{3,Ne}$ state. Another excited state of Neon, $E_{2,Ne}$, is located in between the $E_{1,Ne}$ and $E_{1,Ne}$ states. As this $E_{2,Ne}$ state normally is unoccupied, population inversion between the $E_{2,Ne}$ and $E_{3,Ne}$ states is easily reached. Stimulated emission between these two states results in red photons of 1.96 eV (632.8 nm). Atoms in the $E_{2,Ne}$ state decay towards the ground state by spontaneous emission.

In gas lasers, such as the HeNe laser, the light is reflected on both ends of a tube filled with the gas, while in solid state lasers, this happens in a solid bar. The waves oscillate back and forth between the ends of the cavity, which is why a laser is sometimes referred to as an oscillator (Figure 4.5). Because of this resonance, increase or extinguishment (interference) occurs according to the length of the cavity in function of the laser wavelength. Because all photons are coherent, the phase difference is constant, after reflection on a mirror. For positive interference twice the cavity length needs to equal a multiple of the laser wavelength. In practice, the laser wavelength is much smaller than the cavity. In addition, a certain band tailing occurs in the laser itself, whereby various wavelengths near to each other are emitted. This is called gain bandwidth. In practice there are different cavity modes (longitudinal modes) which fit in the gain bandwidth. The intensity distribution over the laser beam is defined by the design of the cavity.

From the mechanisms, as shown in Figures 4.3 and 4.4, it is clear that a certain stimulus is needed to excite the atoms: the atoms need to be pumped towards a higher energy level. In most cases several of these excited states are considered, and polychromatic light can be used. In these cases a discharge lamp or a flash light is used.

This process is called optical pumping of a laser. Of course the pumping photons must have a shorter wavelength (higher energy) than the laser light. Sometimes, a primary laser (pump laser) is used to excite the atoms in a second laser.

A second mechanism which is often used is electrical pumping, for example in a gas laser or in a diode laser. In a gas laser an electric current is sent through the gas, whereby the molecules are sent to a higher energy state, to start the decay process in that way. With some gas lasers a continuous current is sent through the gas, whereby a continuous laser output is formed. With other gas lasers short electric pulses are sent, whereby a pulsed output is formed.

Semi-conductors lasers (Figure 4.6) work differently, but also use electrons to generate the population inversion. In these cases the inversion is produced between the current carriers (electrons and electron-hole pairs) in the junctions between the various regions in the semi-conductor. The diode consists basically of a p-junction and an n-junction, with the active layer in between. Electrical leads are connected to both junctions, and the diode is positioned in a heat sink, to remove the generated heat. The light emission in the semi-conductor is concentrated in the active layer and both sides of the semi-conductor crystal act as mirrors for the cavity. The semi-conductor chip and the heat sink are mounted in a protective casing and the light, under the form of an elliptical beam, can escape through a window. The semi-conductor chip reflects sufficient light in the crystal to form an efficient stimulation. The semi-conductor crystal is usually polished, to improve efficiency. The main advantage of diode lasers is that a lower electrical current (and voltage) is needed compared to gas lasers.

4.1.3 Lasers for Raman Spectroscopy

In general, the Raman spectrometers which use lasers with a wavelength shorter than 750 nm are dispersive Raman instruments,

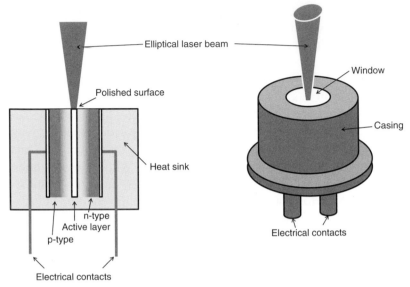

Figure 4.6 Diode laser.

while Fourier-transform (FT) spectrometers are used for longer wavelengths. The reason for this lies with the signal/noise ratio of the used detectors and the sensitivity of the different detectors in specific spectral ranges. In practice, Raman spectroscopy lasers are used, with wavelengths in the UV area (<200 nm) until in the near infrared (1064 nm). A summary of some frequently used laser wavelengths is given in Table 4.1.

QUESTION 4.1

Calculate the wavelength for a (Stokes) Raman band at 3000 cm^{-1}, as detected by using a Nd:YAG laser as well as for a HeNe laser.

Table 4.1 Commonly used lasers for Raman spectroscopy.

Laser type	Frequently used laser wavelengths in Raman spectroscopy (nm)
HeNe laser	632.8
Kr^+ laser	413.1, 647.1
Ar^+ laser	488.0, 514.5
Diode laser	660–880
Nd:YAG laser	1064
Frequency doubled Nd:YAG laser	532

This summary only discusses continuous wave lasers; there are, however, as well applications of pulsed lasers for Raman spectroscopy. Through their extremely high power density serious safety measures are needed, and the sample may be damaged.

For Raman spectroscopic applications a few conditions need to be complied with:

- Frequency stability: we do not want the laser wavelength to change significantly between two measurements.
- Narrow bandwidth: for most applications a narrow laser line (monochromatic) is required, typically smaller than $1\,cm^{-1}$, although for some routine-analyses this is not necessary. The bandwidth of the laser line also affects the resolution of the established spectrum.
- Few sidebands (use of band pass filters) gas lasers emit atomic emission lines of the present gases, while solid state lasers can emit luminescence radiation. This radiation can interfere with the sample and cause fluorescence among others.
- Low divergence: especially when no optical fibres are used, this is an important criterion.

- Good definition of the profile of the laser: this is necessary if the spot size on the sample needs to be well defined (cf. confocality, see later).

QUESTION 4.2

When a laserline with a width of 1 cm^{-1} (Raman shift) is required, what is the corresponding width expressed in nm for a 785 nm diode laser? And for a 632.8 nm HeNe laser? And for a 413.1 nm Kr$^+$ laser?

4.2 Detectors

For Raman spectroscopy, different types of detectors are used. For dispersive instruments we distinguish single-channel detectors and multi-channel detectors, while for Fourier-transform instruments semi-conductor detectors are used (mostly Si, InGaAs or Ge). Different types of detectors are discussed successively.

Taking into account the very weak Raman signals, it is necessary to dispose of sensitive detectors. It is not only important to have a high quantum efficiency (the number of electrons generated per photon), but we also want to have a dark signal as low as possible. This dark signal is formed by spontaneous (thermal) generation of electron–hole pairs in the semi-conductor detector, and involves the introduction of noise (see next chapter). Therefore, these detectors are cooled. As a rule, the longer the wavelength, the less energetic are the photons. Therefore electron–hole pairs are formed less easily and the sensitivity in the infrared region of the electromagnetic spectrum decreases. Detectors that are still sensitive in the near IR only need little energy to create an electron–hole pair and as a consequence a small amount of thermal energy is sufficient to form an electron–hole pair; therefore, the detector needs to be cooled more and the contribution of the detector background increases.

In regards to detectors a few terms are important:

- Quantum efficiency (quantum yield, detector sensitivity): the number of photo electrons generated per photon that reaches the detector.
- Response curve (quantum efficiency curve): graph which shows the quantum efficiency in function of the wavelength (Raman wavenumber).
- Number of channels: for a multi-channel detector, this is the number of signals/wavelengths that can be measured independently from each other. For the often used charge-coupled device (CCD) detectors this is the number of pixels along the wavenumber axis.
- Dark signal (detector background): the average number of electrons generated in the detector, if the detector is not exposed to light.
- Dark noise and readout noise: noise generated because of the dark signal and by the quantisation by digitalisation respectively. (See next chapter.)

4.2.1 Single-Channel Detectors

Historically, photon multiplier tubes were first used in Raman spectroscopy in the 1960s. The advantage of this approach is their relatively high sensitivity compared to the earlier measurement geometries. The disadvantage of these tubes is their rather high background signal (typically $10\,e^-/s$) and their sensitivity to permanent damage: if the detectors are exposed to extremely high light intensities (e.g. laser beam or lights in the room), this could result in permanent damage. Moreover, this approach has the disadvantage that you can only measure at a single wavelength and that the analysis of a complete spectrum needs to be done sequentially, which is time-consuming.

4.2.2 Multi-Channel Detectors

In practice people want to record multiple wavelengths simultaneously. Therefore, usually a multi-channel detector is used. A multi-channel detector is basically a linear array of single-channel detectors. A particular part of the spectrum is projected along the linear axis of the detector, and thus, each channel of the detector corresponds with a different wavelength (or wavenumber). Calibration of this type of spectrometer consists then of assigning a wavenumber to each element of the detector. Linear detectors are often used in small, handheld Raman instruments.

4.2.3 Charge-Coupled Device (CCD) Detectors

Linear detectors often suffer a relatively bad signal to noise ratio. Therefore, in dispersive Raman instruments, another type of multi-channel detector is often used, which consists of a 2-dimensional array of light-sensitive elements: a charge-coupled device (CCD) detector. Usually, the spectrum is projected along one of the dimensions of the CCD, and pixels of the other direction are combined – a method which is named binning. However, some Raman instruments use the full 2-dimensional properties of a CCD, to incorporate spatial information in the analysis (see Chapter 5, Section 5.7 on imaging).

CCD chips usually consist of silica, where photons create electron/hole pairs. In Figure 4.7 a typical response curve for a

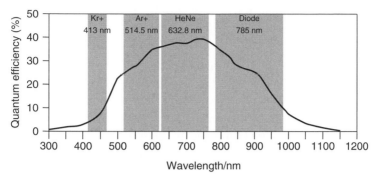

Figure 4.7 Typical response curve of a CCD, with a typical range of a Raman spectrum $(0-3000\,\text{cm}^{-1})$ for different lasers.

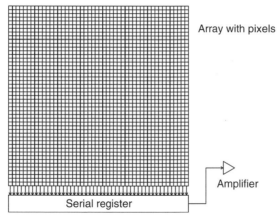

Figure 4.8 Illustration of the design of a CCD-detector.

CCD is shown, with the wavenumber ranges (0 to $3000\,\text{cm}^{-1}$) in relation to various laser wavelengths.

On the chip an electric circuit is printed, consisting of a 2-dimensional matrix of contact points (Figure 4.8), which are kept at a positive potential. Each contact point corresponds with a small area, which is sensitive to photons. These areas can each contain between 10^4 and 10^6 electrons, before they are saturated (full well capacity). The potential drain therefore functions as an integrator which during a period of time accumulates the electrons which are generated by the photons. Because these potential drains are placed in a two-dimensional array, each column corresponds with a specific wavelength. Often the pixels per column are therefore added up (binning), whereby super pixels are formed. Binning is the combination (before (hardware binning) or after (software binning) readout) of various pixels. During read-out, the charges which are stored in the potential drains are converted to digital entities, whereby the gain (sensitivity) of the detector is the proportionality factor between the digital signal and the number of photoelectrons.

We can distinguish between a front-illuminated and a back-illuminated CCD. The front of a CCD is considered as the side where the electrical circuit is printed. A front-illuminated CCD is illuminated from this side. The electric circuit covers approximately half the area and is not photon-sensitive, therefore the maximum quantum efficiency is ca. 50%. Normally the back of a regular front-illuminated CCD can be partially etched (back-thinned), whereby the thickness is only ca. 15 μm. Thus, the electrons which are formed by the irradiation of the back can migrate through the chip to the potential drains at the front of the detector. Because the electric circuit does not cover the CCD now, the quantum efficiency is much higher than with a front-illuminated CCD (typically up to 80–95%). Furthermore, the detector is more sensitive in the UV area, compared with a front-illuminated CCD. This is because, with the making of the electric circuits, people use UV-absorbing coatings (at the circuit side). Back-illuminated CCDs are often covered with an anti-reflective coating. Some typical response curves are shown in Figure 4.9.

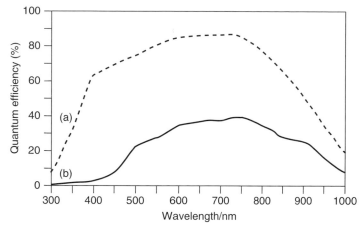

Figure 4.9 Typical response curves for (a) back-illuminated and (b) front-illuminated CCDs.

Although back-illuminated CCDs have a much higher sensitivity than front-illuminated CCDs, there are a few relevant disadvantages:

- Back-illuminated CCDs are more expensive than front-illuminated CCDs.
- Back-illuminated CCDs are thinner and therefore more fragile than front-illuminated CCDs.
- With back-illuminated CCDs light passes through a thin layer of silica and therefore an interference phenomenon can occur. This etaloning effect is reflected in oscillations in the sensitivity and is more pronounced with the use of longer laser wavelengths. Some CCD producers could reduce this effect, but it is best to test the CCD before purchasing it.

Another effect which is used to increase the sensitivity of the detector is the use of a deep depletion CCD. This term refers to the

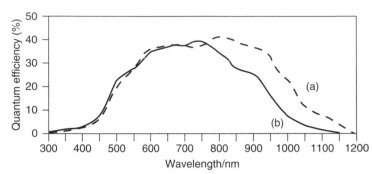

Figure 4.10 Response curve for (a) a typical front-illuminated CCD and (b) a deep-depletion front-illuminated CCD.

use of a doped silica chip, to increase the sensitivity at relatively longer wavelengths (see Figure 4.10).

In conclusion, it is clear that the most appropriate CCD needs to be selected in function of the wavelength area in which you want to work (cf. Figure 4.7).

4.2.4   Semi-conductor Detectors

To avoid fluorescence, it can be useful to use excitation in the infrared region (see Chapter 2). However, as can be seen from the response curves, CCD detectors are not sensitive in this spectral region. Therefore, FT-Raman instruments (with typical wavelength between 1100 and 1700 nm) are usually equipped with semi-conductor detectors with a small band gap (forbidden zone). This band gap should be sufficiently small to allow low energy photons (long wavelengths) to create electron–hole pairs in the semi-conductor. Often indium gallium arsenide (InGaAs) and germanium (Ge) are used. If the forbidden zone is sufficiently small that the low energetic photons can be detected, the detector background will unfortunately increase accordingly, since electron-hole pairs are generated spontaneously. Therefore, these detectors need to be cooled, as the 'thermal' creation of electron–hole pairs is thus reduced. Various factors strongly influence the detection efficiency (e.g. doping and other manufacturing effects); it is therefore usual for Raman manufacturers to select detectors with the highest quantum efficiency or the lowest dark noise from a production batch. Therefore, the instrument sensitivity or the noise can change drastically after the replacement of a detector.

4.3 Filters

A Raman instrument contains a number of filters, each with their own characteristics. In general we can divide the filters into different categories (polarisation filters are not considered here):

- Neutral density filters (grey filters): these absorb the light with a quasi-constant value over the whole wavelength area.
- Long pass filters: These filters reflect the light with a wavelength under a certain limiting value; above this limiting value there is transmission.
- Low pass filters: Light with a wavelength shorter than a certain limiting value will be let through, whereas light with a longer wavelength is reflected.
- Band pass filters: These filters only let light pass of a certain wavelength (within a certain wavelength interval), the other light is reflected.
- Band block filters: These filters block light from a certain wavelength, but let the light through with another wavelength.

An important term when studying filters is the optical density (O.D.). The optical density of a filter is defined as:

$$O.D. = -\log(\%T)$$

where %T is the percentage of the transmitted light. In other words, if a neutral density filter allows 1% of the light to pass, the optical density of that filter is 2:

$$2 = -\log(0.01)$$

The higher the optical density, the more the filter blocks the light.

QUESTION 4.3

What is the optical density of a 10% and a 25% filter? If a given filter has an optical density of 0.30, what percentage of the light is transmitted?

Since the Raman effect is an intrinsically weak effect, it is important that the Rayleigh signal is suppressed. In the past, people used a double or triple monochromator. Nowadays, holographic filters are often used in Raman instruments. These filters consist of a gelatine-based material in between two glass plates, in which a standing wave pattern is generated. We can consider this filter as a pile of thin layers of gelatine with various densities. If this filter is irradiated with light of a suitable wavelength, destructive interference occurs in this filter, whereby the light of this wavelength is absorbed. It goes beyond saying that to get the optimal effect, the wavelength of the standing wave pattern (i.e. the thickness of the layers) is important. Therefore, during production of these filters this standing wave pattern is generated with a laser of (approximately) the same wavelength as the radiation to be blocked. You can vary the path length in the filter by tilting the filter at a certain angle. Thus, it is possible to adjust the precise position (wavelength) of the absorption band (notch) in the spectrum.

A disadvantage of holographic filters is that they degrade. During functioning, part of the laser light is absorbed, and the energy is released as heat. The standing wave pattern tends to blur after heating and thus the spectral quality of the filter is reduced. Dielectric edge filters are also available on the market. They are a bit more expensive than holographic filters, but are said to degrade much more slowly. Dielectric films consist of alternating micrometer-thin layers of transparent materials (e.g. TiO_2, SiO_2) that are sputtered on a glass substrate. Similar to holographic filters, their working principle is interference-based. However, unlike holographic filters, which are a type of band-block filters, dielectric edge filters are long pass filters.

QUESTION 4.4

What is the consequence of using dielectric filters (being long pass filters), when you want to record Stokes and anti-Stokes Raman spectra?

4.4 Dispersion Systems

As outlined in Figures 4.1 and 4.2, all spectrometers need to be able to disperse the light, in function of their wavelength. Light of different wavelengths can be separated either as function of space or in function of time. The first type of spectrometers use the principle of diffraction of light, whereas the second type use Fourier-transformations to obtain the result.

4.4.1 Systems Based on Diffraction of Light

The first type of spectrometer is commonly named 'dispersive instruments'. The most simple diffraction instrument is a prism, where diffraction is based on the wavelength-dependent (angle of) refraction of the light in the prism. However, in modern spectrometers, usually diffraction gratings are used as dispersion system. A diffraction grating can be considered as a linear repetition of reflecting (or sometimes transmitting) elements. The distance between the elements (typically referred to as grooves or lines) is similar to the wavelength of the light that needs to be dispersed. Due to the periodic variation in the diffraction grating, the scattered (or transmitted) light undergoes constructive or destructive interference, depending on the wavelength of the light and the angle under which the light is scattered, and thus a spatial resolution of the light, as a function of its wavelength is obtained. Two main types of gratings are differentiated: a reflection grating and a transmission grating. Dispersive gratings consist of a reflective substrate on which a diffraction grating is superimposed. The incident and diffracted light are on the same side of the grating. Opposite to this, when using a transmission grating the incident light beam and the diffracted light are on different sides of the grating (Figure 4.11).

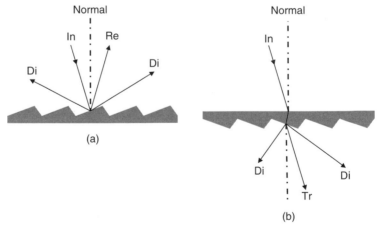

Figure 4.11 Schematic drawing of the lightpaths in (a) a reflection grating and (b) a transmission grating. In: incident beam; Re: reflected beam; Tr: transmitted beam; Di: Diffracted beams.

When monochromatic light is sent under an angle α to a grating, the light is diffracted and under different angles β_n constructive interference can be observed. The angles under which constructive interference is observed are dependent on the angle of incidence, the wavelength of the light and the groove spacing d of the grating. Considering two parallel rays of light, incidenting on a grating (Figure 4.12), constructive interference will only occur when the difference in pathlength between both scattered rays equals an integer number of wavelengths (λ). This is:

$$\text{n.}\ \lambda = d\ (\sin \alpha - \sin \beta) \tag{4.1}$$

This relationship explains why, for monochromatic light, constructive interference occurs under different angles. The integer n is called the order of diffraction. From this equation it can be seen that the zero-order diffraction corresponds with a situation of reflection ($\alpha = \beta$), which is wavelength-independent. This means that under zero-order conditions the grating acts as a mirror.

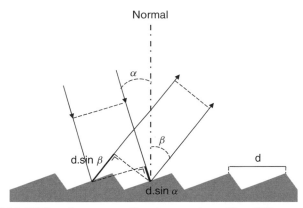

Figure 4.12 Diffraction at a reflective grating. The difference in pathlength between the two rays (in bold), equals d(sin α − sin β).

When changing the angle of incidence α between the grating and the incident beam (e.g. by rotating the grating), one can select the wavelength(range) that reaches the detector. Indeed, by rotating the grating, the angle between the detector and the normal of the grating, a signal is observed for a particular wavelength, if this angle corresponds with the angle of constructive interference β. The spectral resolution and spectral range, which is projected on the detector, are dependent of the groove spacing d.

However, when selecting a grating for a spectrometer, not only the desired spectral resolution and spectral window are of importance. For a given groove spacing, different gratings can have a different efficiency. Indeed, some materials are better suited than others for dispersion in particular wavelengths. Moreover, gratings can be coated and the exact shape of the grooves can be different. Therefore, it is important to study the efficiency curves of the grating when selecting it for a particular application.

4.4.2   Fourier-Transform (FT-) Systems

The central part of an FT-Raman spectrometer is the Michelson interferometer. This set-up (Figure 4.13) consists basically of a beam-splitter and two mirrors, of which one is fixed and the other is translated during the experiment. If we consider a monochromatic light source, which emits light in the interferometer, the light beam is split into two equal parts. 50% of the light is reflected, while 50% of the light passes the semi-transparent mirror. Both of these beams hit a mirror and are sent back to the beam-splitter, where they are combined and sent to the detector. Dependent of the difference in pathlength of these two beams, constructive or destructive interference occurs and the detector registers a signal or not, respectively. When the moving mirror is moved during this experiment, the detector detects, as a function of time (or, if you like, as a function of the position of the moving mirror) a fringe pattern.

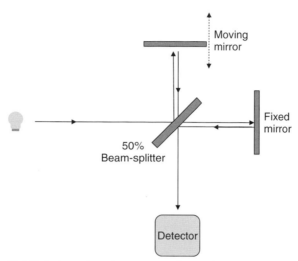

Figure 4.13 Michelson interferometer, the main component of an FT-Raman spectrometer.

During a Raman-experiment, light of different wavelengths (a spectrum) is sent to the interferometer. Each of these wavelengths generates its own fringe pattern and the fringe patterns of all wavelengths that are present in the spectrum are superimposed and detected by the detector. Thus, a so-called interferogram is obtained. Usually, the quality of an interferogram resulting from a single translation of the moving mirror is of poor quality, due to high noise levels. However, to improve the quality of the interferogram, multiple scans (typically a few hundred or thousand) are recorded one after the other and added together. In order to obtain a normal Raman spectrum, a mathematical operation is performed on the interferogram, namely an inverse Fourier-transformation. The spectral resolution that can be obtained with an FT-instrument, is related to the distance the mirror travels per scan. This can be understood by realising that for a given wavelength the number of fringes that is detected is dependent on the distance that the mirror travels. In order to be able to perform a precise deconvolution, sufficient fringes need to be present. It should be noted that, contrary to dispersive instrumentation, there is no trade-off between spectral resolution and spectral coverage. Often, people mention the multiplex-advantage when discussing FT-Raman spectroscopy. By this term it is meant that all wavelengths are determined simultaneously, although using a single-channel detector. In contrast to this, dispersive spectrometers when using a single-channel detector, need to scan all the wavelengths one after the other. Usually, the movements of the moving mirror are checked with a second laser, which allow for excellent frequency precision.

4.5   Components for Transportation of Light

Apart from the filters, Raman instruments contain other optical components (e.g. mirrors, lenses, etc.) to bring the laser light to the sample, and to lead the scattered light to the Raman spectrometer. In general, two approaches exist for this: the use of a set of lenses and mirrors and/or the use of glass fibre optics. In any case the lenses and mirrors need to be equipped with the appropriate coatings to enable the transmission, respectively the reflection as high as possible. For instance, when working in the infrared region, it is of the utmost importance that the optical components are water-free, as this strongly absorbs infrared radiation, which not only causes less intense spectra and increased noise levels, but the heat which results from the absorbed radiation may cause damage to the components.

The transmission of light through glass fibres is based on the total reflection of the light beam, against the walls of the glass fibre. It is generally thought that the greatest losses with use of glass fibre optics take place at the extremes. There are different parameters which can play a role:

- The material of which the glass fibre is made determines the absorption and therefore also the transmission efficiency of the light. Therefore it is necessary to opt for a glass fibre with few -OH functional groups, because they may absorb the light. The fibre may have to be applied with a suitable coating.
- The diameter of the glass fibre determines how much light can be sent through the cable: it is easier to focus the laser beam in a thick glass fibre than in a thin glass fibre. On the other hand it is true that with a thin fibre you can work in single mode, which means that less band tailing will occur and that the laser spot on the sample will spatially be better defined. This is necessary if you want to work in a confocal way.
- The numerical aperture of the fibre determines the maximum solid angle with which the beam is sent into the fibre and under which the beam leaves the fibre.
- The angle under which the fibre is cleaved determines the direction with which the laser beam leaves the fibre.
- The geometry with which the different fibres are placed also determines the result. It is important that there is an overlap as large as possible between the volume irradiated with exciting beam, as the area which is covered by the collection fibres. There are different measurement geometries possible for the probe heads (Figure 4.14).

When using optical fibres one has to consider that the glass fibres themselves can also initiate a Raman signal or a fluorescence signal. Therefore, filters need to be used to avoid or eliminate these signals (Figure 4.15). In the probe head, before hitting the sample, the laser beam passes through a bandpass filter. Thus, the beam is 'cleaned' and possible signals with other wavelengths are removed, in order that only laser light reaches the sample. The back-scattered light passes through a band rejection filter before entering the collection fibre. There the most intense signal, the Rayleigh line, is removed.

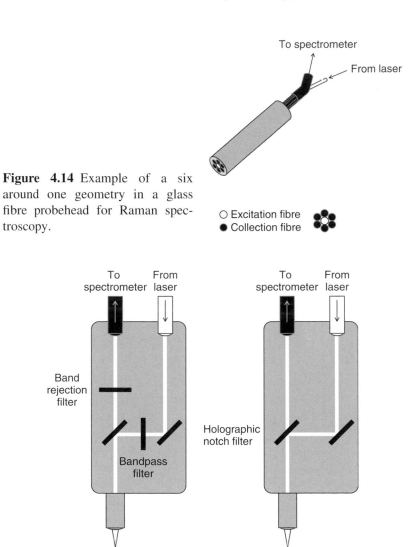

Figure 4.14 Example of a six around one geometry in a glass fibre probehead for Raman spectroscopy.

○ Excitation fibre
● Collection fibre

Figure 4.15 Schematic design of two fibre optics probe heads for Raman spectroscopy.

The Rayleigh line does not yield useful information for Raman spectroscopists, but if not removed it would induce a Raman or fluorescence signal in the collection fibre, which would overwhelm the collected Raman spectrum. A more compact design consists of using a single holographic notch filter (or dielectric filter) that acts as a mirror for the laserline, while light of other wavelengths can pass.

4.6 Sample Chambers and Measurement Probes

For Raman spectroscopy, different types of sample chambers and measurement probes were designed. It is impossible to discuss the details here of all types. Therefore we will limit ourselves to some important, general characteristics of this interface between the sample and the spectrometer. These principles will therefore be illustrated through a few examples.

From the formula for the intensity of the Raman signal, it can be concluded that, in order to obtain a signal as intense as possible, various factors are important: the intensity of the laser beam and the number of molecules that are irradiated. Furthermore, the collection of the Raman diffused light needs to be conducted as efficiently as possible. If we perform macroscopic analyses, then the number of analyte molecules (concentration) is a characteristic of the sample in which the analyst usually does not interfere (unless a pre-concentration on a substrate is done).

With macroscopic analysis of transparent materials, it can be interesting to make the path length in the sample as long as possible: this way the laser can interact with as many molecules as possible. This principle is applied when using gas cells, where they let the laser light go back and forth through a sample room filled with gas.

A similar principle is often used with the analysis of liquids, when part of the sample vessel is covered with a reflecting surface.

The intensity of the laser radiation on the sample needs to be as high as possible, without damaging the sample. Therefore it is necessary that the light is sent as efficiently as possible through the Raman spectrometer. The lens, which focuses the laser light on the sample, needs to have a transmission as high as possible. When selecting the objective lenses for these purposes some characteristics are important.

Firstly, the transmission of the laser light through the objective[2] needs to be as high as possible. Therefore, the objectives need to be made of a suitable type of glass. A typical problem for the use of intense IR lasers is that the lenses and the IR-coating may not contain any water: by absorption the lens warms, whereby small lens deformations may occur and the beam is not well focused. The lenses might be treated with an anti-reflective coating.

In general, there are two types of important image-deformations which can occur to the lenses, namely spherical and chromatic aberration. Spherical aberration occurs with a single lens and is formed by the beams which go through the centre of a lens and are not focussed at the same focal point as beams which are refracted outside the lens axis. It is clear that the larger the solid angle of the lens, the stronger this lens aberration is. Chromatic aberration is formed because the light with different wavelengths is deflected differently by the lens and therefore has a different focal distance. To correct for these aberrations, compound lenses are usually used for microscope lenses: a stack of different lenses, fitted in one housing, tries to correct for these aberrations.

---

[2] Objective lens: here the lens which focuses the laser light on the object is named the objective lens, although this lens does not necessarily have to be connected to a microscope. For some Raman measurements, for instance, telescope lenses are used.

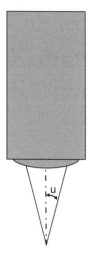

Figure 4.16 Schematic drawing of the aperture cone of an objective lens.

An important parameter, which describes the size of a conical beam of light that can pass through a lens, is the numerical aperture (N.A.). This quantity can be calculated with the formula:

$$\text{N.A.} = n \cdot \sin u \qquad (4.2)$$

with n the refractive index of the medium and u the half top angle of the light cone (Figure 4.16). Therefore, the N.A. is a measure for the amount of light which can be collected with the objective lens. In general, it can be said that for most objectives the N.A. becomes greater at a decreasing focal distance. The refractive index n for air is equal to 1; in practice the N.A. for normal objective is 0.95 at the very most. If higher N.A.'s are required, then another medium needs to be present between the objective lens and the object (e.g. water or immersion oil, which have a higher refractive index). Furthermore, certain objectives are corrected for the use of cover slides: these thin layers of glass have an refractive index which is different from 1, which causes the focal point to be at a different distance than with noncorrected objectives.

Other important characteristics of the objectives are:

- the transmission efficiency: the higher the transmission efficiency, the more light can pass the objective, and the fewer losses occur as a result;
- the enlargement: determines how the laser beam is focused, and therefore which surface is analysed on the object/sample;
- the working distance: the space between the bottom edge of the fitting of the lens and the focal plane;
- the focal distance: the distance between the lens and the focal plane;
- the iris: the diameter of the lens; this one needs to be greater than the diameter of the light beam which is to go through the lens.

One of the often used sample–spectrometer interfaces is a microscope: the laser beam is led into the microscope unit, whereby the microscope objective is used as focusing lens. The same objective lens is used to collect the Raman-diffused light and to send it to the spectrometer. The big advantage hereby is that pieces of sample can be measured with a very small diameter, allowing us to study inhomogeneities.

The lateral resolution is defined according to the Rayleigh criterion, where the resolution δ_p with the illumination of the object with a parallel light source (e.g. laser beam) is given by:

$$\delta_p = 1.22 \cdot \lambda/\text{N.A.} \qquad (4.3)$$

with λ the laser wavelength and N.A. the numerical aperture of the lens. This limit value, however, is not reached in practice. Besides the lateral resolution, the depth resolution also plays an important role. This depth resolution can be greatly improved by confocal measurements. Notice that in what follows, people consider that no significant light refraction nor absorption occurs in the sample, which can seriously change the measured volume.

Opposite to classical microscopy, where the field is illuminated homogenously, with confocal measurements spatial filtering is used, with the use of pinholes or diaphragms. This assembly isolates the light from a specific plane in the sample which coincides with the illuminated part, and eliminates the radiation efficiently coming from other planes which are out of focus. With this set-up it is possible to avoid the Raman signal and the fluorescence contribution of areas which are not in focus. An approximate formula for the maximal obtainable depth resolution is given by:

$$\delta_z \geq \left| \frac{4.4 \cdot n \cdot \lambda}{2\pi \cdot (\text{N.A.})^2} \right| \tag{4.4}$$

An experimental set-up for confocal measurements is shown in Figure 4.17, where it is illustrated how only the radiation of a well-defined depth can pass to the spectrometer. In practice a pinhole or diaphragm is placed in the collected light beam, or the light is led in a glass fibre, where the entrance of the glass fibre functions as the confocal opening.

4.7 Noise in Raman Spectroscopy

Noise is inherent to every spectroscopic technique. With the term 'noise' we mean all possible signals that are detected and that do not contain relevant information. Noise can seriously influence the precision of the obtained results. Typical examples of noise in Raman spectra are fluorescence, spikes and background noise. Spikes are said to be caused by cosmic rays that accidentally hit the detector. They result in sharp peaks in the spectrum (typically only few

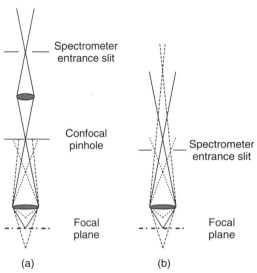

Figure 4.17 Scheme of a confocal (a) and a conventional (b) microscope, coupled to a spectrometer.

pixels wide) and are not reproducible: when recording the same spectrum again, no signal is detected on this spectral position.

The total noise level (σ) contains contributions of different factors, of which some are inherent on the sample (σ_s), some depend on the instrumentation (σ_i) and some on the signal processing (σ_p):

$$\sigma = \sqrt{(\sigma_s^2 + \sigma_i^2 + \sigma_p^2)} \tag{4.5}$$

4.7.1 Noise Originating from the Sample: σ_s

It is a basic rule in spectroscopy that all spectroscopic processes are subject to a certain uncertainty. The uncertainty on a given spectroscopic signal is fundamental and is, by laws of statistics, given by:

$$\sigma_s = \sqrt{S} \qquad (4.6)$$

where S is the signal intensity. This type of noise is commonly referred to as shot noise, and cannot be avoided. Since, for dispersive spectrometers, the total measured signal is proportional to the time measured, it can be said that the shot noise is proportional to the square root of the time. When using an FT-instrument, noise is proportional to the square root of the number of scans.

It should be noted that the shot noise is proportional to the signal intensity. In many cases the signal is composed of different portions, such as the Raman signal of the analyte (in which we are interested), the Raman signal of other components in the sample (matrix), fluorescence detected by the detector, stray light entering in the detector, etc. All these different signals contribute to the total shot noise level. Often it is possible to correct for certain signals, by performing mathematical operations, such as spectral subtraction of a blank to correct for matrix influences or mathematical subtraction of a polynomial to correct for broadband fluorescence signals. However, it is important to understand that it is not possible to correct for the contribution of the shot noise, caused by these interferences.

Since all these signals are proportional to time, one can expect that the signal to noise ratio of a spectrum can always be improved by measuring for a longer time. However, this effect is minimal, since the major contribution to the noise can be due to the background noise.

An important distinction needs to be made between the use of dispersive Raman instruments and the use of FT-spectrometers. In a dispersive instrument, the shot noise on a certain band position corresponds to the square root of the signal, as recorded at that specific pixel. Indeed, the signal on a certain position is not influenced by the signal on another pixel. When recording FT-spectra, contributions of all wavelengths hit the detector simultaneously. Therefore, the shot noise in an FT-instrument is not only determined by the shot noise of the signal, but is dependent on the total detected signal. This means that a small Raman band at a certain position can be overwhelmed by the noise of the fluorescence in another region of the spectrum.

4.7.2 Noise Originating from the Raman Instrument: σ_i

All components in a Raman instrument can contribute in a way to the Raman spectrum. For the sake of good spectrometer engineering an instrument is designed with low spectrometer noise. If certain components (e.g. glass fibres) give rise to extra signals, shot noise (see previous paragraph) is associated with this contribution.

Detectors are also a source of noise. This is due to the detector dark signal: a certain number of electron–hole pairs are spontaneously generated, even when no photons are present. This thermal generation can be reduced by cooling the detector. This dark signal contributes to the detector shot noise and is also proportional to the square root of measurement time. Detector noise is usually the limit for FT-Raman experiments.

Lasers also contribute to the instrument generated noise level in a spectrum. Some lasers typically do not emit light with a constant intensity, but tend to flicker (depending on the type of laser). FT-instruments are hardly sensitive to this effect, since all wavenumbers are measured simultaneously. Also dispersive spectrometers with a fixed grating are rather insensitive to flicker noise, since the whole spectral range is measured at once. However, when the diffraction grating is tilted and different parts of the spectrum are recorded in sequence – or a single-channel detector is used – the spectrum might suffer some influence of laser flickering.

4.7.3 Noise Originating from the Signal Processing: σ_p

The processing of the electrical signals is also subject to noise. An important type of noise in this context is detector readout noise, which is generated when converting the analogue signal (a number of electrons) to a digital value in the computer. It originates from, amongst others, the amplifiers and other electronical components and the analogue-to-digital-convertor. The detector readout noise is not proportional to the measurement time, and detector read-out noise is usually only encountered when performing very short measurements or when measuring very weak signals (e.g. gases or monolayers of molecules).

One should as well be aware that mathematical operations on spectra (spectral post-processing) can have an influence on the noise levels. For instance, multiplying the spectral values with a constant (without deviation) changes the absolute noise level, but the signal to noise ratio remains constant. However, adding (or subtracting) a constant (e.g. during baseline subtraction) does not change the absolute noise level, but the signal to noise ratio is influenced (as the signal intensity changes). The rules that apply in these cases are commonly named 'error propagation'.

4.8 Summary

In this chapter, the general structure of Raman instrumentation has been described and each component has been discussed in detail, so that the reader can understand their use. Special attention has been given to the laser, as monochromatic light source and to the CCD detectors and filters. We have discussed in detail the fundamental differences between a dispersive instrument and a Fourier-transform (FT-) Raman spectrometer. Moreover, by now, the reader should also be able to understand the different sources of noise in a Raman spectrum and see how a dispersive and an FT-Raman spectrum are diffcrently affected by noise.

Further Reading

Gardiner, D.J. and Graves, P.R. (eds), *Practical Raman Spectroscopy*, Springer, New York, 1988.

Graselli J.G., Bulkin B.J. (eds), *Analytical Raman Spectroscopy*, John Wiley & Sons, Inc., New York, 1991.

Laserna, J.J (ed.), *Modern Techniques in Raman Spectroscopy*, John Wiley & Sons Ltd, Chichester, UK, 1996.

McCreery, R.L., *Raman Spectroscopy for Chemical Analysis*, John Wiley & Sons, Inc., New York, USA, 2000.

Chapter 5
Raman Spectroscopy in Daily Lab-life

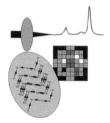

Learning Objectives

- To know the principles of calibrating a Raman spectrometer
- To understand the working principle of digital filters
- To appreciate how to interpret Raman spectra of organic and inorganic molecules
- To recognise the pitfalls for quantitative Raman spectroscopy
- To understand the working principles of spectral searching algorithms
- To know the advantages and disadvantages of Raman mapping versus Raman imaging

Raman spectroscopy is becoming increasingly popular as an analytical tool in chemical laboratories – and beyond. Indeed, the absence of complex sample preparation procedures makes the technique attractive, also for nonspecialist users. However, there

Practical Raman Spectroscopy: An Introduction, First Edition. Peter Vandenabeele.
© 2013 John Wiley & Sons, Ltd. Published 2013 by John Wiley & Sons, Ltd.

are some pitfalls, which may lead to the idea that the technique 'does not work well' or which may lead to wrong interpretations, and thus it is important to have support of experienced Raman users. In this chapter, we try to give some examples on common routines in daily life in a Raman spectroscopy laboratory.

5.1   Calibration of a Raman Spectrometer

Calibration of the equipment seems an obvious action in an analytical research lab. However, depending on the required degree of precision and accuracy, the calibration procedure should be more or less stringent. For instance, in many cases, when performing only qualitative analysis, no intensity calibration is required. On the other hand, when looking for minute differences or band shifts between spectra, for instance by using chemometrical techniques, good calibration of the spectrometer is necessary.

Basically, there is calibration of the X-axis in the spectrum on the one hand, while there is intensity calibration (Y-axis calibration) on the other hand. For qualitative research, the first type is the most important, whereas for quantitative both aspects of calibration are of high importance. X-axis calibration refers to making sure that all pixels correspond with the wavenumbers they should refer to, whereas intensity calibration is a procedure to relate the spectrometer response to the intensity of the Raman scattering. In the following paragraphs, we will discuss both types of calibration subsequently.

5.1.1   X-Axis Calibration

By performing an X-axis calibration, one tries to annotate a meaningful number to the different pixels of the CCD detector in a dispersive instrument (Figure 5.1). This can be on the one hand by assigning Raman wavenumbers (in cm^{-1}) to the different pixel numbers, or on the other hand by relating the pixel numbers to absolute wavelengths (in nm). Instrument manufacturers usually recommend calibrating the instrument by performing an interpolation of the laser line ($0\,cm^{-1}$) and the intense band of silicon ($512\,cm^{-1}$). The advantage of this approach is in its easiness: it is not difficult to focus the laser beam on a piece of silicon wafer and in a couple of seconds a spectrum is recorded. The disadvantage of this approach is that this is a two-point measurement and that a standard calibration curve (usually a line) is drawn through these points. From a statistical point of view, this approach is not robust at all: if something goes wrong with the measurement of the exact band position, the whole spectrum will be miscalibrated. Moreover, when someone is analysing Raman bands at high wavenumbers, these are quite a bit off the calibrated region – and extrapolation is always tricky.

Another approach consists of recording multiple Raman spectra of different compounds, with well-known Raman band positions. Thus, someone can make a calibration, based on a larger number of Raman shifts. Moreover, one can select reference products with bands in the spectral region of interest.

If the user is also interested in monitoring the stability of the laser – i.e. measuring the laser wavelength, then it is important to record a spectrum, that allows to relate the pixels with absolute wavelengths, expressed in nm. Therefore, usually sources with well-known and narrow spectral lines are used, such as the spectral emission lines of gases, in certain lamps. Typically, neon lamps or argon tubes are used. When selecting the tube, it is important to have as many emission lines as possible in the spectral region of the laser line.

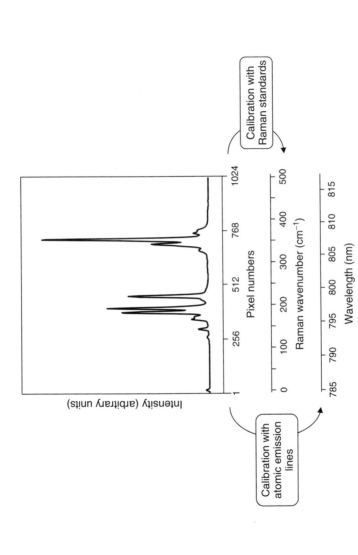

Figure 5.1 Overview of X-axis calibration, with the Raman spectrum of the mineral realgar (As_4S_4) as an example.

5.1.2 Y-Axis Calibration

A spectrometer is not evenly sensitive in all regions of the electromagnetic spectrum. Indeed, as can be seen from the response functions of CCD-detectors, they are not sensitive in certain spectral regions, such as the infrared region (see Chapter 4). Moreover, each pixel may have a slightly different sensitivity and also the dark signal of the detector needs to be corrected. In addition, all optical components in the light path may selectively absorb light with specific wavelengths, or may cause fluorescence that selectively causes a higher background signal at specific wavelengths. A good Y-axis calibration corrects for all these interferences.

The Y-axis calibration consists of a two-stage process. On the one hand, the detector dark signal is recorded. This is the signal that is generated by the detector if no light hits the detector. Therefore, the lens and laser are closed and the detector is set open, to record a spectrum during the same time as normally is measured. The recorded spectrum is a dark spectrum and should be subtracted from the final spectrum. On the other hand, a white light spectrum should be recorded. This should take the wavelength-dependent detector sensitivity into account, as well as the influence of all optical components. Ideally, an intense light source should be placed in front of the spectrometer entrance (e.g. in front of the probe head or under the objective), so that it completely fills the aperture of the objective. However, usually it is quite complicated to position this lamp in such a way, and therefore, a diffusor is sometimes used, that reflects the light homogeneously. Ideally, a lamp is used that has exactly the same intensity over all wavelengths (perfect white light) and the recorded spectrum reflects the sensitivity of each pixel. In practice, however, no such ideal white light exists, and a calibrated lamp with a well-known emission pattern should be used. When such a lamp is purchased, the emission spectrum is delivered together with it – and thus, the wavelength-dependent emission intensity is known. In order to be able to use it properly, it is necessary to perform a wavelength calibration.

Discussion Topic 5.1

Explain why is it necessary to perform a wavelength calibration to be able to use a calibrated white light source.

Answer:

The information provided with the calibrated lamp is basically an emission spectrum: the intensity of the lamp as a function of its wavelength. From the Raman spectrometer, we obtain information in the form of a raw spectrum: measured intensity as a function of pixel number. The Y-axis calibration has as a purpose to define the correction factors needed to let the measured intensity correspond with the intensity of the lamp, at that particular wavelength. In order to know what pixel corresponds with the wavelength (expressed in nm), it is necessary to perform a wavelength calibration, for instance by using a neon lamp.

When performing the two above-mentioned steps (i.e. correction for the dark signal and white light calibration), the recorded spectrum can be presented with correct relative intensities. However, this result does not yield an absolute calibration. Indeed, as was seen in the first chapter, the absolute intensity of the Raman signal is dependent on different factors, such as instrumental factors and the intrinsic intensity of the Raman scattering molecule. We have corrected some instrumental factors already, but some factors are not yet included. These include the measurement geometry and the laser intensity. In order to be able to relate concentrations of the analyte molecule with the measured Raman intensity, it is necessary to analyse different samples with known analyte concentrations and to make a calibration line that plots the (corrected) intensities as a function of the analyte concentrations. If the standards have a similar matrix as the concentration of the unknown, this approach is also able to correct for self-absorption. If good-quality lasers are used, the laser intensity should not change significantly between different measurements.

The idea of working with a dilution series is quite appealing, when working with samples in solution or in the gas phase. However, for solids, this is not straightforward, especially when combined with micro-Raman spectroscopy. Indeed, sample inhomogeneity should be several orders of magnitudes smaller than the measured volume.

INTERMEZZO 5.1 MICRO-RAMAN SPECTROSCOPY IN MICROBIOLOGY

Micro-Raman spectroscopy is a very versatile analytical technique. It is one of the few approaches that allow us to obtain molecular information on a micrometer-scale. Indeed, it is possible to record spectra with a spectral footprint (this is the area that is measured) of less than 1 μm^2 – a feature that is extremely welcome in the research of very small organisms, like bacteria.

Bacteria grow on different media, and may cause different diseases or may cause food to degrade. In the food industry it is therefore important to detect possible contamination as soon as possible. In health care, it may be important to identify the bacteria fast, in order to give the patient the most suitable treatment. Traditionally, samples are taken and incubated in petri dishes for 24 hours – or in some cases even longer. Subsequently, different techniques are used to identify the strains, such as microscopy with appropriate staining. Under certain conditions, Raman spectroscopy can be an alternative and faster approach. Indeed, as this technique is able to pick up a molecular signal in a small volume, the technique does not need a large amount of biological material: a micro-colony is sufficient for the analysis, which seriously reduces the culturing time. The recorded Raman spectrum contains features of all molecules present in the sampling volume. As the over-all composition of each bacterial species is different, the obtained signature can be sufficiently specific to identify the species.

However, differences between different species are minimal and often chemometrical techniques are used to differentiate between spectra. Minimal differences between spectra, due to different concentrations of one of the molecules in the bacteria, can be reflected by different band intensities or slight band shifts (in the case of overlapping bands). These small shifts can only be detected if the instrument was properly calibrated. Especially if databases of reference spectra are made, it is very important to correct for instrument instability, by appropriate wavenumber and intensity calibration. Moreover, in microbiology research, it is important to work under standardised conditions. Indeed, factors like the medium and the culturing conditions (time, temperature) can seriously influence the shape of the Raman spectrum.

Further Reading

De Gelder, J., De Gussem, K., Vandenabeele, P., Vancanneyt, M., De Vos, P., Moens, L., *Anal. Chim. Acta* **603**/3 (2007) 167–75.
Hutsebaut, D., Vandenabeele, P., Moens, L., *Analyst* **8**/130 (2005), 1204–14.
Hutsebaut, D., Maquelin, K., De Vos, P., Vandenabeele, P., Moens, L., Puppels, G.J., *Anal. Chem.* **76**/21 (2004), 6274–81.

5.2   Raman Spectral Post-processing

Raman spectra are often processed before they are presented. Besides calibration, other forms of spectral post-processing comprise the use of digital filters, scaling and curve fitting. The latter type is often used for baseline corrections.

5.2.1 *Digital Filters*

Digital filters exist of many types and can have many purposes. Often these are so called window functions.

The working principle of a simple window function is illustrated in Figure 5.2. Firstly, the type of function should be defined. In this case, the arithmetic mean is chosen, but other options, like a weighted average, a polynomial fit or a derivative can be selected, depending on the aim of the filter. Secondly, the window size should be defined. In its basic form, the number of pixels in a window is an

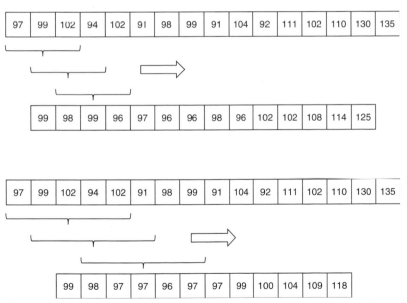

Figure 5.2 Working principle of a window function: example of a 3 pixel and a 5 pixel smoothing function.

odd number. Depending on the complexity of the function, there is a minimal window size. Indeed, if the function consists of taking the fifth order polynomial, then a window size of 3 pixels is insufficient. When the window function is run, the function is applied over the range of the window (in the upper case of Figure 5.2: take the mean over 3 subsequent values and round the result). The result of the function is assigned to the central pixel of the window. Finally, the window moves to the next position and the process is repeated: the result is calculated and assigned to the new central pixel. The window moves until the end of the spectrum is reached. Note that, if a window of 2N+1 pixels is selected, the size of the resulting spectrum is reduced with 2N pixels.

QUESTION 5.1

What would be the outcome of the spectrum in Figure 5.2 if a 3-cell window function is applied with a weighted average function, where the central pixel has a the double weight compared to the other pixels?

In Figure 5.3 the effect of the above-mentioned window function is illustrated with a Raman spectrum. It can be clearly seen that the effect of noise is reduced. The larger the selected window, the more the noise level is reduced. However, small features may as well be eliminated. Moreover, band broadening occurs and in some cases even slight band shifts may be observed. Therefore, where possible, the use of these functions should be avoided, and if used, this should clearly be mentioned, with inclusion of all parameters.

5.2.2 Scaling

The absolute intensity of a Raman spectrum is dependent on different factors, including some quite experimental factors that

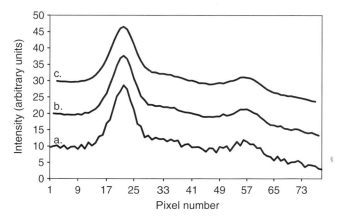

Figure 5.3 Window function: example of a smoothing function. (a) original spectrum; (b) 3 pixel smoothing function; (c) 5 pixel smoothing function. Spectrum c and b are shifted by resp. 10 and 20 intensity units, for better viewing.

are difficult to control. Often, people are interested in relative intensities of the different bands, and to compare between different spectra, scaling can be of use. Different types of scaling can be considered. The most simple form of scaling is a linear scaling, where the maximum intensity (I_{max}) in the spectrum is set to 1, and the lowest intensity (I_{min}) is set to 0. The corresponding equation to transform the original intensity ($I_{original}$) to a new scaled intensity I_{scaled} is:

$$I_{scaled} = (I_{original} - I_{min})/(I_{max} - I_{min})$$

This approach has the advantage that the whole spectrum varies between +1 and 0. A variant on this approach consists of dividing the original intensity by the sum of the original intensities at all spectral positions:

$$I_{scaled} = I_{original}/ (\Sigma\ I_{original})$$

Other algorithms are often used when chemometrical calculations are involved. Then, sometimes the average intensity of the spectrum is set to 0, where 1 standard deviation corresponds with 1:

$$I_{scaled} = (I_{original} - \mu_I)/SD_I.$$

In this equation, μ_I is the mean intensity of the original spectrum, and SD_I is the standard deviation of the intensities in the spectrum. This scaling transformation is often named Standard Normal Variate (SNV) and was first proposed by Barnes *et al.* (R.J. Barnes, M.S. Dhanoa and S.J. Lister, *Appl. Spectrosc.* **43**, 772–7 (1989)).

When performing multivariate analysis based on a collection of Raman spectra, it might be useful or even necessary to perform one of these scalings on the different variables. In this case, the mean and standard deviation are not taken over an entire spectrum, but over the intensities corresponding with a particular wavenumber in the different spectra in the dataset. In other words, in a data matrix, the scaling is not performed on the rows, but rather on the columns.

5.2.3   Deconvolution

Deconvolution is the mathematical procedure where a spectrum is resolved in different bands. Firstly, the type of curve needs to be selected. Indeed, often one can select different types of bell-shape, that more or less match with the shape of the Raman band. Typically, a Lorentzian of Gaussian function can be fitted. Theoretically, an ideal band should be of a Loretzian shape, but in practice different components may cause band broadening. Therefore, a Voigt-function is based on a Lorentzian shape, with additionally some Gaussian band broadening.

Once the type of curve is selected, other parameters may determine the outcome of the deconvolution. Each band has its own bandwidth and intensity; depending on the software package, these parameters should be set by the user, or they can be determined automatically.

One may think that it is easy to determine the band position, and in the case of a single Raman band this may be the case. However, when several bands are overlapping and shoulders are present, it is impossible to determine the exact band position. Some software packages are, by deconvolution, able to produce a list of band positions, which may create the illusion of objectivity. Unfortunately, this is only an illusion as many different (preset) parameters influence the outcome of the analysis. These include the number of bands, their bandwidths, etc. The software tries to optimise all these settings for the different fitted curves, in order to have the smallest residual as possible. The danger of applying this procedure is that all parameters need to be optimised at once and the uncertainty on one band influences the result of another band.

5.2.4 Baseline Corrections

In a Raman spectrum, apart from the Raman bands, other features are present in the spectrum, such as fluorescence or background radiation. For some applications, it may be useful to eliminate these broad features from the spectrum. A baseline correction is the ideal tool for this. A large number of approaches exist for this problem. Two main approaches can be distinguished: on the one hand for some approaches the user needs to select a certain number of points while on the other hand there are also fully automated approaches.

In the first case the user selects a series of points which he assumes are on the baseline. Then, depending on the type of baseline correction, the software draws straight lines between these points. Thus, a baseline is created, which is subtracted from the original spectrum. Alternatively, the software can calculate a curve (e.g. exponential curve or a polynomial), which is used as baseline. It is obvious that the manual approaches are quite time-consuming,

especially if a large dataset is involved. Moreover, the selection
of the points that define the baseline is quite stringent and may
strongly influence the outcome. Sometimes, the shape of the bands
may look strange. There are even software packages that allow the
users to select points that are not on the curve, and thus allow
the user to artificially create or enhance bands. On the other hand,
the manual approach (and especially the first approach where short
intersects are defined) deals relatively well with sudden changes in
the spectral background, such as the edge caused by a Rayleigh
band-block filter.

Alternatively, different automatic approaches are developed.
Usually, these approaches are based on iterative processes. One
way is illustrated in Figure 5.4. A polynomial is fitted to the entire
spectrum (including the Raman bands). In Figure 5.4a, a 2nd
order function was fitted. In the next phase, a baseline is defined
containing all points of the spectrum with their intensity lower
than the defined polynomial. For the other points, the values of the
polynomial are included. In the next iteration, a new polynomial is
calculated, to fit to this baseline. This procedure is repeated. Two
approaches exist: either you repeat the procedure for a predefined
number of times, or you repeat it until no further improvement
is obtained (i.e. if the difference between the newly defined
polynomial with the previous one is smaller than a limiting value).

When chemometrical processes are performed, sometimes one
can take derivatives, for instance by using a window function.
Indeed, the width of the window is much smaller than the broad
fluorescence features. Therefore, they can be filtered. The resulting
spectrum does not look like a normal spectrum, as for each Raman
band the derivative is taken – but this may enhance the outcome of
the chemometrical calculations.

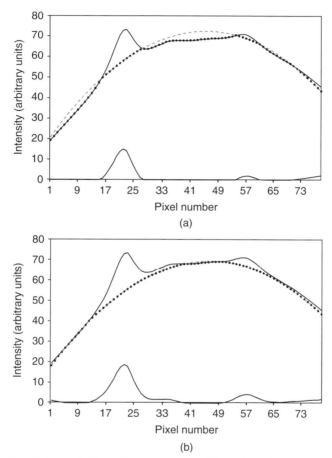

Figure 5.4 Polynomial baseline correction. The original spectrum is plotted in a solid black line. The fitted 2nd order polynomial is in grey dashes. The selected points for the baseline are marked with dots: (a) first iteration and (b) seventh iteration.

5.3   Interpretation of Raman Spectra of Organic Molecules

Interpretation of Raman spectra can be quite complicated, especially if no other information but the Raman spectra is available on the samples. When using the term 'spectral interpretation', usually the assignments of different vibrations to the Raman bands is meant. This may help in the identification of the molecules. In generally, two types of spectral interpretation can be distinguished: positive and negative. A spectrum should be interpreted by looking at the bands that are present (positive interpretation) as well as which bands that are not present (negative interpretation). The idea of positive interpretation is quite obvious, but negative interpretation might seem less evident.

We will explain the approach of negative interpretation with some examples. Most organic molecules contain C-H groups. It is clear that, when one expects an organic molecule in the sample, the analyst will examine the spectral region around $3000\,cm^{-1}$. In this region, the ν(C-H) stretch vibration is expected. Especially, the Raman band of ν(C-H) of aliphatic molecules is expected below $3000\,cm^{-1}$ (more specifically between ca. 2750 and $3000\,cm^{-1}$), whereas the ν(C-H) band for aromatic molecules, alkenes and alkynes is expected above $3000\,cm^{-1}$. The absence of these bands may indicate that no organic molecule is present – although there are some organic molecules that don't contain any C-H groups (e.g. C_6Cl_6). Moreover, when δ(C-H) bending vibrations are assigned, ν(C-H) stretching vibrations should be present as well – good Raman spectroscopists examine the whole spectrum and try to assign different bands simultaneously.

Organic molecules usually yield quite complex Raman spectra, with many different Raman bands. The bands can shift due to other groups present in the mixture, or in the analyte molecule: electron donors and electron acceptors can influence the exact band

Table 5.1 Typical Raman band positions for some organic functional groups.

$3350-3300\,\mathrm{cm}^{-1}$	$\nu(\equiv\text{C-H})$	C-H stretch of alkynes
$3150-3000\,\mathrm{cm}^{-1}$	$\nu(\text{C-H})$	C-H stretch of aromates
$3000-2750\,\mathrm{cm}^{-1}$	$\nu(\text{C-H})$	C-H stretch of alkanes
$2300-2050\,\mathrm{cm}^{-1}$	$\nu(\text{C}\equiv\text{N})$	C≡N stretch of nitriles
$1870-1650\,\mathrm{cm}^{-1}$	$\nu(\text{C=O})$	C=O carbonyl stretch
$1750-1450\,\mathrm{cm}^{-1}$	$\nu(\text{C=C})$	C=C stretch of alkenes
$1670-1630\,\mathrm{cm}^{-1}$	$\nu(\text{CONH})$	Amide I vibration
$1350-1250\,\mathrm{cm}^{-1}$	$\delta(\text{CONH})$	Amide III vibration
$1150-950\,\mathrm{cm}^{-1}$	$\nu(\text{C-C})$	C-C stretch of alkanes
$1050-950\,\mathrm{cm}^{-1}$	$\nu(\text{CC})$	CC ring breathing of aromates

positions. An overview of some typical Raman band positions for functional groups of organic molecules are shown in Table 5.1.

INTERMEZZO 5.2 RAMAN SPECTROSCOPY IN A FORENSICS LABORATORY

Raman spectroscopy – and especially micro-Raman spectroscopy – is a tool that is increasingly more often used in a forensics context. The ability of the technique to obtain a spectrum of small samples easily is especially promising. The technique has, for example, been used to identify minute volumes of explosives or gunshot residues.

Another application consists of the analysis of automotive paints. This can be of importance in identifying suspect vehicles after a hit-and-run car accident. Some traces of the escaped car can remain on the other object. Analysing the different layers in the automotive paint and their pigments may be indicative for the make and model of the car. Automotive paints consist of different layers, such as clear-coat and base-coat. Minute samples need to be embedded in a resin, so that the paint flake can be investigated from its side and the different layers can be seen. Due to the excellent lateral resolution of micro-Raman spectroscopy, each layer can be studied separately.

Raman spectroscopy is also frequently used in a forensics lab when it comes to fast analysis of found drugs. The speed of analysis and the possibility of recording a molecular spectrum through a plastic can may be achieved simply by focussing the laser beam through the plastic. Thus, it is not necessary to open sealed bags to be able to identify their content. When using portable spectrometers, this can easily be done on location.

Further Reading

De Gelder, J., Vandenabeele, P., Govaert, F., Moens, L., *J. Raman Spectrosc.* **11**/36 (2005), 1059–67.
Hargreaves, M.D., Page, K., Munshi, T., Tomsett, R., Lynch, G., Edwards, H.G.M., *J. Raman Spectrosc.* **7**/39 (2008), 873–80.

In the next paragraphs we will discuss some Raman spectra of various organic molecules, thus providing the basic principles of Raman spectral interpretation of alkanes, cycloalkanes, alkenes and aromates. In the following figures, a series of Raman spectra of organic molecules are shown, so that the reader can follow the description of typical band positions by comparing them with the Raman spectra of selected organic molecules. In some cases, for easy comparing, spectra are repeated in different figures.

However, it has to be taken into account that that some of these spectra are recorded on different Raman spectrometers, by using different lasers, namely a green laser (532 nm) and an infrared laser (785 nm). In Figure 5.5 the Raman spectra of two compounds, n-hexane and cyclohexane, as recorded on the

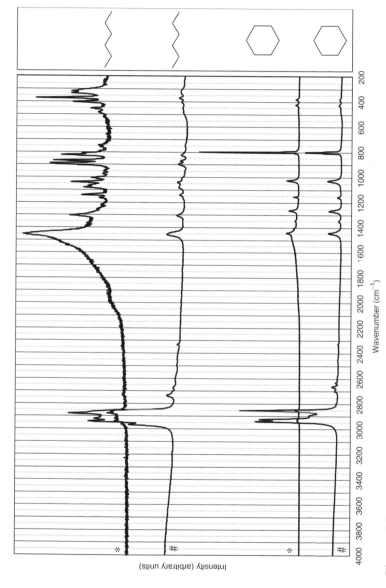

Figure 5.5 Raman spectra of n-hexane and cyclohexane. (*) Spectra recorded by using a red 785 nm laser. (#) Spectra recorded with a green 532 nm laser.

different spectrometers, are presented. Basically, band positions
are independent of the laser used. However, the appearance of the
spectrum is totally different: relative band intensities have shifted.
The main reason for this is that the spectral response curve is
different for both instruments. Indeed, as seen previously, classical
CCD detectors are insensitive in the infrared region (see Section
4.2.3). As a consequence, the relative intensity of the bands around
$3000\,cm^{-1}$, compared to the other bands of the spectra, is much
lower when using a 785 nm laser. Moreover, when using the 532 nm
laser, a small Raman band seems to be present at ca. $2330\,cm^{-1}$.
However, this band is not observed when using the 785 nm laser.
Consequently, this is not a Raman band, but rather an artefact,
caused by stitching two parts of the Raman spectrum – the whole
spectrum is not recorded at once, but in two parts, that afterwards
are stitched together. Therefore, in the next figures, we will omit
this Raman band. The spectra recorded with the near infrared laser
(785 nm) tend to exhibit some tailing towards the high wavenumber
edge, especially for the intense Raman bands around $1450\,cm^{-1}$.
This can be attributed to the optics (objective lenses, magnification)
used for the spectral recording. The spectra, as presented here, are
not pre-processed (e.g. baseline corrected, or dark correction).

5.3.1   *Raman Spectra of Alkanes and Cycloalkanes*

In Figure 5.6, Raman spectra of a series of alkanes and cycloalkanes
are presented. These organic molecules consist only of C-C and
C-H bonds. Therefore, we basically expect to see bands attributed
to ν(C-H) and ν(C-C) stretching vibrations, and to δ(C-H) and
δ(C-C) bending vibrations. In alkanes and cycloalkanes, all carbon
atoms are sp^3 hybridised. ν(C-H) Stretching vibrations are typically
observed at wavenumbers around $3000\,cm^{-1}$. These relative high
wavenumbers can easily be explained by remembering that the
band position is determined by the energy of the first vibrational

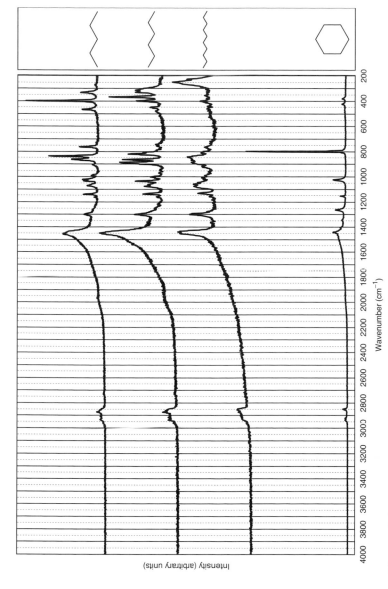

Figure 5.6 Raman spectra of alkanes and cycloalkanes. From top to bottom: n-pentane, n-hexane, n-decane, cyclohexane. Spectra recorded by using a red 785 nm laser.

energy level, hence by the reduced mass and the bond strength of the vibration (see Section 1.3). As H-atoms are very light, this easily explains the relative high wavenumbers of these Raman bands. Raman band position for the ν(C-H) stretching vibration of alkanes and cycloalkanes (i.e. hydrogen atoms bound on single bonded C-C) are typically found between 2750 and 3000 cm^{-1}. Typically, the antisymmetrical stretching vibration is found at slightly higher wavenumbers, compared to the symmetrical stretching vibration. Moreover, ν(C-H)$_{CH3}$ stretching vibrations for a methyl (-CH$_3$) group are also found at higher wavenumbers, compared to the ν(C-H)$_{CH2}$ stretching vibration of ethyl (-CH$_2$-) groups. An overview of typical band positions of ν(C-H) stretching vibrations is given in Table 5.2.

Next to the stretching vibrations, also δ(C-H) bending vibrations can be observed. The δ(C-H) bending vibrations of CH$_2$ groups may occur in two directions compared to molecular backbone and can be symmetrical or antisymmetrical. These vibrations were given specific names, as pointed out in Figure 5.7. Moreover, in some cases, several -CH$_2$- groups in the alkane structure, can vibrate in phase, e.g. in the case of in-phase twisting, a vibration that is Raman active.

In Figure 5.6, a series of Raman bands is observed between ca. 2950 and 2840 cm^{-1}. These can be attributed to the ν(C-H)

Table 5.2 An overview of typical band positions of ν(C-H) vibrations.

Vibration	Functional group	Spectral range (cm^{-1})
ν(C-H)$_{sym}$	-CH$_3$	2860–2880
ν(C-H)$_{asym}$	-CH$_3$	2950–2900
ν(C-H)$_{sym}$	-CH$_2$-	2840–2870
ν(C-H)$_{asym}$	-CH$_2$-	2910–2940

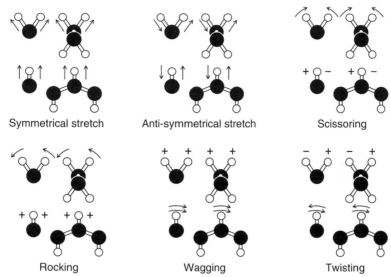

Figure 5.7 CH stretching and bending vibrations of CH_2 groups in alkanes. For each vibrational mode, the CH_2 group is presented as single group (left) and in combination with other CH_2 groups in an alkane (right). For each vibrational mode, the CH_2 group is drawn once in frontal view (top) once in side view (bottom). The arrows and + and − signs indicate the movement during the first half of the vibration. + and − signs indicate movements in and out of the plane, respectively.

stretching vibrations in the alkanes. Although it is difficult to assign specific Raman bands in this region, due to the spectral overlap and degeneration, in general, the following sequence is observed, starting at the high wavenumber edge: $\nu(\text{C-H})_{\text{CH3,antisym}}$, $\nu(\text{C-H})_{\text{CH2,antisym}}$, $\nu(\text{C-H})_{\text{CH3,sym}}$, $\nu(\text{C-H})_{\text{CH2,sym}}$. Where the bands in this region tend to overlap in the case of linear alkanes, in the case of cyclohexane, the Raman bands seem much better resolved. As cyclohexane does not contain -CH_3 groups, these Raman bands are not observed in its Raman spectrum. In more strained ring systems (e.g. smaller rings), the ν(C-H) stretching vibrations shift to higher wavenumbers. It is also clear that the relative contribution of the CH_3 groups, compared to the CH_2 groups, is higher for short chain alkanes. Around 1450 cm^{-1}, a relative intense Raman band is observed in the alkanes. This band can be assigned to an overlap of δ(C-H)$_{CH3,antisym}$ and δ(C-H)$_{CH2,sciss}$ bending vibrations (obviously, for cyclohexane, this band is only attributed to δ(C-H)$_{CH2}$). Note that the symmetrical ν(C-H)$_{CH3,sym}$ stretching vibration, between ca. 1350 and 1400 cm^{-1}, is usually of very weak intensity in the spectra of alkanes. The Raman band at ca. 1300 cm^{-1} can be assigned to the in-phase δ(C-H)$_{CH2\ twist,in-phase}$ vibration. Typically, the relative intensity of this band becomes higher for longer n-alkanes.

In alkanes, apart from the C-H vibrations, also C-C vibrations can be observed. These vibrations are typically marked as skeletal vibrations, involving ν(C-C) stretching and δ(C-C) bending modes of the carbon chain. These Raman bands are typically observed below 1300 cm^{-1}. Typically, for longer chain alkanes, these bands are observed as a series of overlapping bands. In general terms, the ν(C-C) stretching modes are observed between ca. 1300 and 850 cm^{-1}. However, when discussing long aliphatic chains, it is sometimes difficult to make a clear distinction between stretching and bending vibrations. Typically, skeletal vibrations are considered as a combination between ν(C-C) stretching and δ(C-C-C) deformations. Their band positions are in general dependent on the chain length of the alkanes. Some of these vibrations can be observed as a Raman band between ca. 835 cm^{-1} for *n*-butane, ca. 870 cm^{-1} for *n*-pentane and between ca. 850 and 905 cm^{-1} for longer *n*-alkanes. The chain expansion is observed in the lower wavenumber range. This is a bending vibration, where all C-C-C angles are changed in phase, resulting in an over-all expansion and contraction of the aliphatic backbone. These Raman bands are typically observed in the range between ca. 400 cm^{-1} for *n*-pentane, ca. 370 cm^{-1} for

n-hexane to ca. $250\,\mathrm{cm}^{-1}$ for n-decane, and even lower wavenumbers for longer alkanes (e.g. ca. $70\,\mathrm{cm}^{-1}$ for $C_{35}H_{72}$). The exact band position, as well as the bandwidth, are dependent on the aggregation state of the sample, liquid or solid.

The Raman spectrum of cycloalkanes is usually dominated by a very intense band, such as the band at $802\,\mathrm{cm}^{-1}$ in the Raman spectrum of cyclohexane. This band is assigned to the ring breathing vibration: this can be considered as an in-phase symmetrical stretching vibration ν(C-C)$_{\mathrm{ring\ breathing}}$. However, the band position of ring breathing vibrations is subject to change, due to ring strain for instance in smaller ring systems or when the structure is substituted, as seen in the spectrum of methyl-cyclohexane (Fig. 5.8).

When comparing the structure of 2,2,4-trimethylpentane (*i*-octane) with n-pentane, it is clear that the first contains an isopropyl (-CH(CH$_3$)$_2$) group as well as a *t*-butyl (-C(CH$_3$)$_3$) group. These groups are reflected in the Raman spectrum (Figure 5.8). The most prominent band in the Raman spectrum is due to the symmetrical ν(C-C$_4$)$_{\mathrm{sym}}$ stretching vibration of the *t*-butyl group, at ca. $745\,\mathrm{cm}^{-1}$. Other, but generally weaker bands, typical for CC$_4$ groups are present around 1250 and $1200\,\mathrm{cm}^{-1}$. The *i*-propyl group gives rise to a symmetrical ν(C-C$_3$)$_{\mathrm{sym}}$ stretching vibration around $830\,\mathrm{cm}^{-1}$.

5.3.2 Raman Spectra of Alkenes and Cycloalkenes

When comparing the ν(C-H) stretching region in the Raman spectrum of cyclohexene with the spectrum of cyclohexane (Figure 5.9), some clear changes can be observed. In molecules containing C=C-H functionalities, usually Raman bands are observed at positions higher than $3000\,\mathrm{cm}^{-1}$. Moreover, ν(C=C) stretching vibrations are typically observed in the region between ca. 1800 and $1630\,\mathrm{cm}^{-1}$. The band positions may shift, depending on the substituents on the C=C double bond: electron donors or acceptors, conjugation, etc.

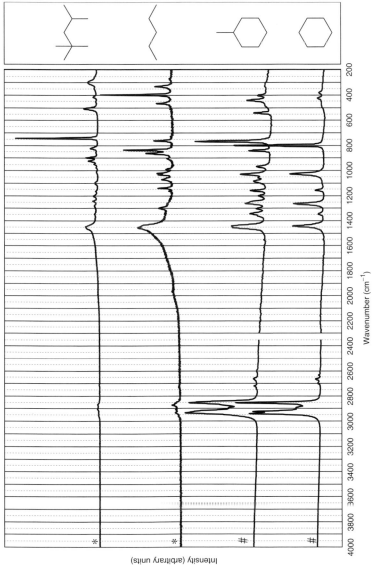

Figure 5.8 Raman spectra of branched alkanes, compared to unbranched alkanes. From top to bottom: iso-octane (2,2,4-trimethylpentane), n-pentane, methyl-cyclohexane, cyclohexane. (*) Spectra recorded by using a red 785 nm laser. (#) Spectra recorded with a green 532 nm laser (2 independently scaled wavenumber ranges).

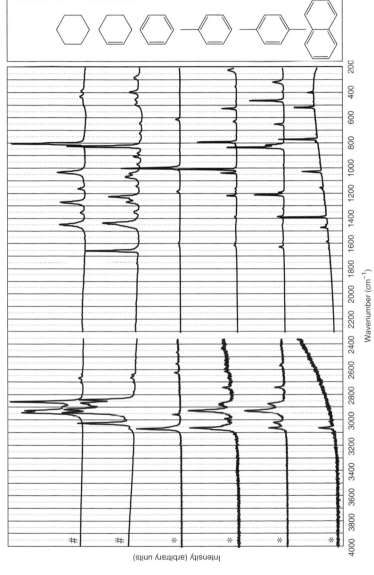

Figure 5.9 Raman spectra of cycloalkenes and aromates, compared to cyclohexane. From top to bottom: cyclohexane, cyclohexene, benzene, toluene, p-xylene, naphthalene. (*) Spectra recorded by using a red 785 nm laser (2 independently scaled wavenumber ranges). (#) Spectra recorded with a green 532 nm laser (2 independently scaled wavenumber ranges).

Table 5.3 An overview of typical band positions of $\nu(C=C)$ and $\delta(C-H)$ vibrations of alkenes.

	$\nu(C=C)$	$\delta(C-H)$
1,1-dialkyl (vinylidine)	$1770-1810\,\mathrm{cm}^{-1}$	$1400-1420\,\mathrm{cm}^{-1}$ $\delta(CH_2)_{scis}$
trans	$1665-1680\,\mathrm{cm}^{-1}$	$1300-1325\,\mathrm{cm}^{-1}$ $\delta(C\text{-}H)_{rock}$
cis	$1630-1665\,\mathrm{cm}^{-1}$	$1250-1270\,\mathrm{cm}^{-1}$ $\delta(C\text{-}H)_{rock}$

Moreover, as can be seen from table 5.3, the band positions of $\nu(C=C)$ stretching vibrations and of $\delta(C-H)$ bending vibrations are dependent on the spatial geometry, and thus, it is possible to discriminate between cis- and trans- configurations. Generally speaking, in unstressed ring structures (like cyclohexene, these Raman bands can be found on similar positions as in the spectra of linear alkenes.

5.3.3 Raman Spectra of Aromates

Benzene molecules are flat hexagonal ring structures, and each carbon–carbon bond is equal in strength and can be considered as 'a bond and a half'. As all C-atoms are sp^2 hybridised, the ring structure is flat, and opposite to the structure of cyclohexane (which is in chair or boat conformation), the molecule has a hexagonal symmetry. This has quite some consequences for the vibrational spectra. A mono-substituted benzene ring has 30 vibrational modes, of which some are Raman active, and some are infrared active. Symmetry properties seriously influence these vibrational modes. Figure 5.10 shows some typical ring vibrations in a benzene molecule. In order to stress the symmetry of the vibrations, some zones of the benzene molecules are shaded: CC bonds that are in the same shaded zone

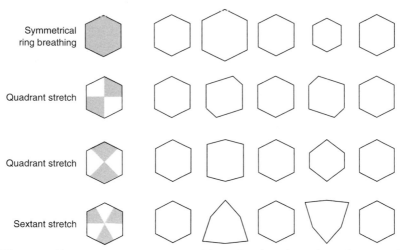

Figure 5.10 Some typical ring stretching vibrations in a planar hexagonal ring structure. Bonds in equally shaded areas act in a similar manner.

vibrate in phase (they stretch and shorten simultaneously), while CC bonds in the opposite coloured zone react inversely.

ν(C-H)$_{Arom}$ Stretching vibrations of benzene derivates are typically observed between 3000 and 3100 cm^{-1}. When comparing the spectra of toluene and xylene with the spectrum of benzene, in the ν(C-H) stretching region, clearly the ν(C-H)$_{CH3}$ methyl stretching bands can be observed. Between ca. 1580 and 1650 cm^{-1}, two Raman bands are observed in the spectra of benzene molecules and substituted benzene molecules. These are assigned to the so-called benzene quadrant stretch vibrations. Typical for unsubstituted and mono-substituted benzene molecules is the very strong Raman band around 1000 cm^{-1}. This band is attributed to a ring breathing vibration. This band is intense for unsubstituted benzene rings, as well as for benzene rings with meta substituents.

In general, the presence or absence of certain Raman bands is symmetry related. Moreover, some typical ring vibrations are more affected by substituents than others. In Figure 5.11 a diagram is given that allows us to determine the position of the substituents on a benzene ring. It must be noted that some band positions might shift, depending on the electrical properties of the substituent: if the group has electron donor or acceptor properties, this might influence the exact band positions. Another remarkable fact in the spectrum of toluene and p-xylene is the presence of a Raman band around 1380 cm^{-1}. This band can be attributed to the symmetrical δ(C-H)$_{CH_3}$ methyl bending vibration. Normally, this band is very weak to hardly observable. In a methyl-substituted benzene functionality (toluene), this band can be observed, due to a hyper-conjugation effect: one of the C-H groups is well-aligned with the π-system of the benzene ring, thus resulting in some Raman enhancement of this particular Raman band.

It is clear that more complex molecules give rise to more complex Raman spectra and are more difficult to interpret. Often, in aromates, heteroatoms like N or O are present in the ring structure. Obviously, this has some implications on the spectra, although often

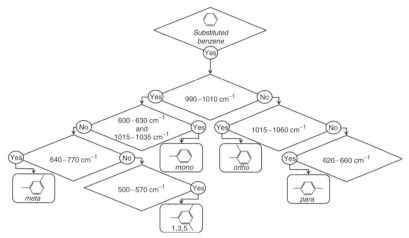

Figure 5.11 The interpretation of Raman spectra of substituted benzene molecules.

similar vibrations (e.g. quadrant stretch, ring breathing, etc.) can be observed.

5.4 Interpretation of Raman Spectra of Inorganic Molecules

In general, inorganic molecules provide simpler Raman spectra than organic molecules: often the number of Raman bands is lower and, especially for crystalline solids, the bands are narrower. On the other hand, the interpretation may sometimes be hampered as the appearance of the spectrum may change, depending on the crystal orientation. As inorganic molecules usually consist of heavier atoms than organic molecules, their Raman bands are usually situated at lower wavenumbers.

As the Raman effect requires a change in polarisability of a molecular bond, ionbonds do not yield a good Raman spectrum. However, ions consisting of multiple atoms, like the sulphate ion

(SO_4^{2-}), may yield a Raman spectrum as the bonds between the sulphur and oxygen atoms are covalent in nature. The number of bonds observed depends on the symmetry of the ion, but the symmetry of the cavity may influence this. Moreover, the Raman band position is not only dependent on the constituting atoms (in this case S and O), but the counter ion or the presence of crystal water may also cause serious shifts.

Discussion Topic 5.2

Why does the counter ion influence the Raman band position?

Answer:

Raman band positions reflect the molecular vibrational energy levels. The position of these energy levels are influenced by two factors, the relative mass of the constituting atoms and the force constant of the bond. Different ions will attract the electrons at different rates and thus, they will influence the force constant of the bond, resulting in a shift in the Raman band position. (see also § 1.1-C, Eq. 1.23).

In Table 5.4 some typical band positions for inorganic groups are given. The shifts, depending on the counter ion, can be quite strong, typically up to several hundreds of cm^{-1}, seriously complicating the band assignments for totally unknown molecules.

INTERMEZZO 5.3 PIGMENT ANALYSIS WITH RAMAN SPECTROSCOPY

Raman spectroscopy has frequently been applied for the analysis of art objects. Special attention has gone to the analysis of pigments. By definition, pigments are colorants that are (unlike dyes) insoluble in their binding medium, so they remain present as particles. Antique pigments are mostly inorganic in nature. The binding medium acts as a glue to stick the pigment grains to each other and on the surface. Thus, a paint film is formed.

The analysis of pigments can reveal important information on the art object. Apart from the fundamental interest in the materials and technique of the old masters, the identification can help in dating the artefact. Indeed, the history of the use of certain pigments is well-known. Identifying pigments about which we know when they were discovered or when they came into vogue or when they disappeared from the artists' palettes can help in dating the artwork. Obviously, care has to be taken not to analyse former restorations or retouches. The identification of anachronisms in the pigment use can be a clue in detecting forgeries. Indeed, when we find pigments that date from a period when the artist was already deceased, one can easily pinpoint a forgery. One can even go a step further and compare the pigments used on the suspect painting with the pigments on other, contemporary paintings of the same artist. Often artists used only a limited number of pigments and if contradictions are found between the pigments that the artist usually used and the pigments on the suspect work, this might be an indication of forgery. Finally, pigment analysis can also be of help in solving specific questions from conservators or restorers.

Further Reading

Vandenabeele, P., Von Bohlen, A., Moens, L., Klockenkämper, R., Joukes, F., Dewispelaere, G., *Anal. Lett.* **15**/33 (2000), 3315–32.

Vandenabeele, P., Bode, S., Alonso, A., Moens, L., *Spectrochim. Acta A* **10**/61 (2005), 2349–56.

Vandenabeele, P., Edwards, H.G.M., Moens, L., *Chem. Rev.* **3**/107 (2007), 675–86.

Table 5.4   Typical Raman band positions for some inorganic functional groups. (After: G. Socrates, *Infrared and Raman Characteristic Group Frequencies – Tables and Charts*, 3rd Edition, John Wiley & Sons Ltd, Chichester 2001.)

CO_3^{2-}	1530–1320 cm^{-1} (m) 1100–1020 cm^{-1} (m-vs) 745–670 cm^{-1} (w)	SO_4^{2-}	1200–1140 cm^{-1} (m-s) 1130–108 cm^{-1} (m-s) 1065–955 cm^{-1} (s) 680–580 cm^{-1} (m-s) 530–405 cm^{-1} (m-s)
PO_4^{3-}	1180–1000 cm^{-1} (s) 1000–900 cm^{-1} (s) 580–540 cm^{-1} (m-w) 415–380 cm^{-1} (m-w)	SO_3^{2-}	1010–900 cm^{-1} (s) 660–615 cm^{-1} (m) 495–450 cm^{-1} (m-s)
NO_3-	1520–1280 cm^{-1} (m-w) 1070–1015 cm^{-1} (s) 860–800 cm^{-1} (m-s) 770–700 cm^{-1} (m-w)	ClO_4-	1170–1040 cm^{-1} (w) 955–930 cm^{-1} (s) 630–620 cm^{-1} (m-w) 490–420 cm^{-1} (m-s)
NO_2-	1400–1300 cm^{-1} (s) 1285–1185 cm^{-1} (m-w) 860–800 cm^{-1} (m-s)	ClO_3-	1100–900 cm^{-1} (m) 980–910 cm^{-1} (s) 630–615 cm^{-1} (w) 510–480 cm^{-1} (m-s)

5.5   Quantitative Aspects of Raman Spectroscopy

In Chapter 1, it was already described that the intensity of a Raman signal is proportional to the analyte concentration in the sample, thus forming the basis for quantitative analysis. Although this may seem straightforward, in practice there are often some drawbacks.

Firstly, matrix effects may occur. Indeed, absorption is one factor that can influence the results. The technique has been tested to measure concentrations in emulsions, with changing success rates.

Sample inhomogenity may influence the results. Matrix effects can partially be overcome by measuring a series of samples with different concentrations of the analyte in a similar matrix as in the unknown sample. The obtained calibration line can be used to estimate the concentration in the unknown sample. When complex matrices are present, standard addition might be a more suitable approach. In this case, a known amount of analyte is added to the sample. By studying the difference in signal intensity between the spectrum of the unknown sample and the spectrum of the unknown with the spiked analyte, it is possible to determine the concentration of analyte in the unknown sample.

A second drawback of the approach is instrument stability. As the measurements of the unknown and the samples for the calibration line cannot be recorded simultaneously, variations in instrumental stability may affect the results. Indeed, if for instance there are fluctuations in laser intensity, these will be reflected in the intensity of the Raman bands in the spectrum and thus influence the accuracy of the results. Moreover, it is not only fluctuations in the intensity that may hamper the measurements, but also fluctuations in Raman band position, which, due to changes in temperature or grating position, will reflect in changing band intensities. Although one always measures the Raman band at the same position of the X-axis (wavenumber), shifts may mean that the bands are not always measured at the maximum intensity of the Raman band. Another factor that may influence the outcome of the measurements is the exact positioning/focussing of the sample in the laser beam. For quantitative measurements often the results obtained by Fourier-transform instruments are of better quality than those obtained by dispersive measurements. This is partially due to the higher stability of the instrument (due to internal wavelength calibration) and to the larger aperture of the spectrometer.

Finally, one has to consider that this approach of making a calibration line is quite straightforward for solutions. When working with solids, however, it is not so easy to make mixtures. Powder mixtures are one solution, but especially when recording spectra by using micro-Raman spectroscopy, sample inhomogenity at microscale level might disturb the outcome of the measurements. In that case many different measurements should be made on different positions or the sample can be rotated to record an average spectrum over a larger area. The disadvantage of this approach is that the spatial resolution is lost. Also, when working with complex mixtures in very small volumes (e.g. when trying to analyse biomolecules in the lumen of a cell), it is practically impossible to create a calibration line, as changing the concentration of the analyte also changes the composition of the matrix.

INTERMEZZO 5.4 DETECTION OF COUNTERFEIT MEDICINES WITH RAMAN SPECTROSCOPY

Counterfeit medicines are a problem in developing countries as well as in developed countries. These fakes may harm the people who take them in a direct or an indirect way. In a direct way the substances included in the counterfeit products may be harmful as such. In a more indirect way, they are harmful as the people think that they are treated or protected, but in the counterfeits there might be no active ingredient or only a low concentration. Moreover, counterfeiting causes serious economical damage, due to the loss of income as well as a reduced trust in the quality of the product and the brand. Therefore, it is important that counterfeit products can be detected and Raman spectroscopy can be an interesting tool to analyse these products. This approach is relatively fast and noninvasive and products can be investigated through blister packages or through glass flasks, simply by focussing the laser beam through the packing. Moreover, small portable instruments become available, so that on-site inspection is easily performed. By using automated spectral searching algorithms, the identification of the products can be automated, so that the technique becomes also accessible for nonspecialist users.

This technique was, for instance, applied for the analysis of artesunate antimalarial tablets. These tablets were purchased on local markets and investigated by Raman spectroscopy. The technique was able to differentiate between genuine and counterfeit tablets. It was possible to identify tablets containing starch, chalk or paracetamol instead of the medicine – and which were clearly counterfeit tablets.

In another research project, Viagra® tablets were examined: these tablets were purchased mostly at nonofficial distribution points. Many counterfeits could easily be detected. Viagra® is a well-known medicine for erectile dysfunction, produced by Pfizer. This company takes many precautions against fraudulent products, including, amongst others, the use of holograms on the packing and well-defined shapes of the tablets. Therefore, most of the counterfeit tablets can be identified by carefully looking at the products and comparing them with the original product.

Further Reading

de Veij, M., Vandenabeele, P., Hall, K.A., Fernandez, F.M., Green, M.D., White, N.J., Dondorp, A.J., Newton, P.N., Moens, L., *J. Raman Spectrosc.* **38** (2007), 181–7.
de Veij, M., Deneckere, A., Vandenabeele, P., de Kaste, D., Moens, L., *J. Pharmac. Biomed. Anal.* **46** (2008), 303–9.

5.6 Fingerprinting and Spectral Searching Algorithms

In many Raman spectroscopic applications, analyte molecules are identified in an unknown sample, not by interpreting the Raman spectrum (identification of the Raman bands), but by comparing the spectrum of the unknown with spectra from a large collection of reference spectra. This approach is named 'fingerprinting'. Databases with reference spectra are available for a broad range of samples and applications. Some spectral databases are sold along with Raman instruments. Unfortunately, the value of these databases is in practice often limited. The spectra are usually recorded on an instrument with different characteristics than the instrument on hand (contributions of different optical components, response curves of the detectors, different spectral resolution, different laser wavelength, etc.) and transferability of spectral databases to different Raman instruments is an issue. Moreover, the origin of the reference products and the spectral recording conditions (e.g. laser power on the sample) should be well-documented. In practice, if possible, it is advisable to record your own database, which exactly corresponds with the field of research of the user.

Small differences (e.g. due to a different fluorescence background) between the spectrum of the unknown and the spectra in the reference database, can easily be taken into account by the experienced user. Manually browsing through the reference spectra is quite time-consuming, especially when large databases are involved. Unfortunately, for computers it is not so easy to eliminate possible interferences or differences between the reference spectra and the spectrum of the unknown.

Many software packages contain an option to search automatically through a database. Usually, the number of options that can be set are limited. In reality, one can imagine of numerous slightly different approaches to look through databases and many options in the software are thus preset. A first consideration is the selection of an appropriate measure of similarity. In chemometrics, different similarity measures are commonly used. Secondly, one can choose to use some technique to avoid interferences from the fluorescence background. Fluorescence is not always significant for the analyte as such, as it can as well be caused by the matrix. In other words, the broadband fluorescence does not necessarily contain analytical information on the analyte molecule. The broad features can be eliminated by performing the search on the (first or second) derivatives of the spectra, instead of a search on the original spectra. Taking a derivative (for instance by using a Savitsky-Golay window function) eliminates broad features. Scaling is another factor that needs to be considered in the searching algorithm.

Finally, one has to consider different chemometrical approaches that can be used to evaluate the spectra. Different procedures have been proposed, including complex approaches, such as artificial neural networks or fuzzy logic algorithms. All different approaches have their own advantages and drawbacks. One of the difficulties is how they deal with mixtures.

INTERMEZZO 5.5 EXOBIOLOGY: RAMAN SPECTROSCOPY HELPS THE SEARCH FOR LIFE ON MARS

The question whether there is life outside the earth is very intriguing, from different points of view. One possible place where ESA and NASA are looking for alien life is our neighbouring planet Mars. Raman spectroscopy is one of the techniques that are selected to be included for a future mission with a rover to Mars. Exobiology is the research area that is concerned with the search for extinct or extent life outside the earth, and all different aspects that are associated with this, including the development of techniques and protocols to optimise this search for life.

One aspect of this research is the definition of life. It is very unlikely that higher forms of life will be detected. However, what one should look for is signatures of extinct or extent life. These are so-called bio-markers: molecules or patterns that are uniquely produced by living organisms. One group of molecules that might fit in this definition are carotenoids, which are relatively easily detected by using resonance Raman spectroscopy and a green laser (e.g. a frequency doubled Nd:YAG laser at 532 nm).

Astrobiological research studies as well organisms on earth, that are able to survive in harsh conditions. It is quite likely that organisms develop similar strategies as these extremophiles to survive harsh conditions on Mars (such as extreme temperature differences between day and night, high level of UV radiation, etc.). One of these strategies, used by endoliths, is to position themselves inside rocks, just below the surface. There they receive sufficient UV radiation to be able to perform photosynthesis, but on the other hand, they are protected to harsh UV radiation that can cause damage. Carotenoids and scytonemin are, amongst others, also used in cells as protection against UV-radiation.

The development of optimal strategy of analysis – that can be performed by a robotic system without interference from an analyst on Earth (communication between Mars and Earth is time-consuming) – is another aspect of this research. As the weight of the instruments is crucial, the use of small battery-operated Raman instrumentation is tested under different conditions. Moreover, decision-making algorithms about where to measure are vital. One has to consider that a rover on Mars is travelling slowly and that the area that can be examined during the lifetime of the rover is within a very short distance of the landing place. An alternative approach to overcome these problems is the use of remote measurements, by using a pulsed laser and a gated detector, which allow spectra to be recorded over large distances. However, a major drawback of current instrumentation is that the power generators for pulsed lasers are relatively heavy.

Further Reading

Jehlička, J., Edwards, H.G.M., Culka, A., *Philos. Trans. R. Soc. A-Math. Phys. Eng. Sci.* **368**/1922 (2010a), 3109–25.

Jehlička, J., Vandenabeele, P., Edwards, H.G.M., Culka, A., Capoun, T., *Anal. Bioanal. Chem.* **7**/397 (2010b), 2753–60.
Villar, S.E.J. and Edwards, H.G.M., *Anal. Bioanal. Chem.* **1**/384 (2006), 100–113.

5.7 Raman Mapping and Imaging

The principle of Raman mapping is quite simple: a sequence of micro-Raman spectra is recorded and between two spectra the position of the laser spot relative to the sample is changed. Information on the exact position of the spectra is stored and afterwards these spectra can be recalculated to form a map. This basic principle is quite simple. In practice, usually the sample is moved by using a motorised XYZ-stage. However, especially when using mobile equipment, the fibre-optics probehead can be moved instead of the object.

When discussing Raman mapping and imaging, firstly we should define the difference between imaging and mapping (Figure 5.12). Mapping is the process that we just described where full spectra are recorded, one after the other. Imaging is a technique where a larger area is illuminated (using a defocused laser beam) and where filters are used to obtain an image of the area. By adjusting the filters, one can try to select a specific Raman band – although in practice the spectral bandwidth is broader than a single Raman band. The information that thus is obtained is the intensity of the light that passes through the filters, as a function of position. Opposite to this, when performing mapping experiments, full spectra are obtained, and one can decide to map, for instance, intensities of a specific band, intensity ratios, band positions, bandwidths, etc. It is even possible to perform chemometrical studies (e.g. principal component analysis) and plot the scores as a function of position.

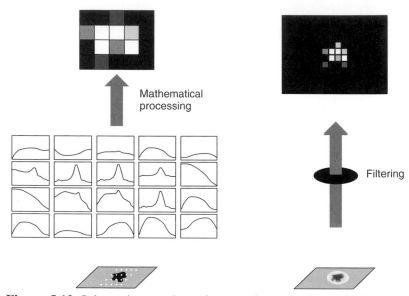

Figure 5.12 Schematic overview of a mapping experiment (left) versus an imaging experiment (right).

Discussion Topic 5.3

How does fluorescence interfere during an imaging experiment? And what happens when performing a mapping experiment?

Answer:

Fluorescence irradiated by the sample can be considered as photons with a specific wavelength that are emitted. In an imaging experiment, the filters are not able to discriminate between photons originating from the Raman effect and fluorescence photons with a similar wavelength. Therefore, the obtained image provides information on the over-all intensity (Raman + fluorescence) as a function of position. In the case of a mapping experiment, one obtains full spectra, and thus a mathematical operation can be performed to eliminate the contribution of the broad fluorescence features. Therefore, the resulting map might look different from the image. For instance, one can perform a baseline correction on all spectra, before constructing the Raman map. In this case, the intensity which is plotted, does not contain a contribution of the fluorescence. Only the noise, caused by the fluorescence cannot be avoided.

A disadvantage of the use of mapping, is that it is quite time-consuming. For a 10*10 Raman map, 100 spectra need to be recorded. Moreover, time is required for positioning of the sample. Mapping works the best when flat samples are involved – thus no problems with focussing are expected. A critical point is that the system should be sufficiently stable, so that no intensity differences (due to for instance fluctuations in laser intensity) are observed during the mapping. In some cases, the instrument uses lenses that focus the laser beam not as a point, but as a line. The projected image on the CCD detector does not exist of a single row, but it contains a whole zone on the detector. The software can then split the obtained image in different rows, so that it appears like the simultaneous measurement of different points in a row. Thus, the mapping can be speeded up. There is even a company which is able to move the stage while the spectrum is recorded. As a consequence, during the measurement, the projection of the wavenumbers on the different pixels of the CCD changes as well, but by correlating the movements of the stage with a mathematical algorithm the original spectra can be reconstructed. Fast mapping procedure is the consequence.

INTERMEZZO 5.6 DIRECT ANALYSIS OF PRECIOUS ART OBJECTS

When analysing art objects, one should always try to reduce the (risk on) damage/obtained information ratio. Reducing the sample size is one way to obtain this, maximising the information that can be obtained (e.g. by using multiple techniques on the same micro-samples) is another approach. Performing direct analysis is a good answer to this dilemma. Raman spectroscopy is well-suited for this approach. Indeed, if low power laser beams are used, no damage to the art object occurs. Moreover, mobile instrumentation can quite easily be transported towards the art object, and in some cases even mapping is an option.

In general, manufacturers put limited attention to the positioning equipment. However, this is of the utmost importance. The stands need to be sufficiently stable, but on the other hand they need to be sufficiently flexible, to allow the user to adapt the set-up to the practical circumstances on hand. Indeed, during a measurement campaign often unexpected situations are encountered. For instance, when analysing wall paintings in an antique Egyptian tomb, electrical power was provided by means of a power generator and the instrument had to be earthed to an archaeological survey nail. As Raman spectroscopic measurements are usually made in a dark laboratory, the doors of the Egyptian tomb were kept closed, resulting in heat staying inside the tomb. On other occasions, for instance when measuring wall paintings on the ceiling of Antwerp Cathedral, the measurements were performed at night, to avoid background radiation. When performing measurements while on a scaffolding, stability is an issue: moving analysts may cause the laser beam to move over several millimetres so that the area may be in or out focus.

When performing measurements in public space, like the exhibition hall of a museum, safety of the visitors is an issue. The audience has to be informed about what measurements take place, but care has to be taken to avoid scattering of the laser. Visitors may not be able to stare into the laser beam. Other important limitations towards in situ investigations are time constraints. Indeed, when working on a (micro-) sample, researchers can always go back to the sample and redo the investigations to double check. When access is given to precious artworks, time is usually limited and the measurements have to be performed during that particular occasion. Therefore, the analyst must be able to make a quick evaluation of the results, to decide on the spot whether extra or longer measurements are needed.

Further Reading

Deneckere, A., Schudel, W., Van Bos, M., Wouters, H., Bergmans, A., Vandenabeele, P., L. Moens, *Spectrochim. Acta A* **2**/75 (2010), 511–19.
Vandenabeele, P., J. Tate, L. Moens, *Anal. Bioanal. Chem.* **3**/387 (2007), 813–19.
Vandenabeele, P., Garcia-Moreno, R., Mathis, F., Leterme, K., Van Elslande, E., Hocquet, F.-P., Rakkaa, S., Laboury, D., Moens, L., Strivay, D., Hartwig, M., *Spectrochim. Acta A* **3**/73 (2009), 546–52.

5.8 Combination with Other Techniques

In an analytical laboratory, often different techniques are combined to answer a question. Raman spectroscopy is in that case not used as a stand-alone technique, but rather as a tool to provide one part of the puzzle. Depending on the questions on hand, different combinations of complementary techniques are made.

One combination that is frequently used is that of Raman spectroscopy with infrared spectroscopy. This combination is extremely fruitful when it comes to structure elucidation. Both techniques provide molecular information about the sample, but when considering the selection rules, vibrations that give rise to a change in polarisation are infrared active, while Raman spectroscopy requires a change in polarisability. Bands of centrosymmetrical molecules are mutually exclusive, but more generally, it is often seen that bands that are very intense in Raman spectroscopy are weak in infrared spectroscopy and vice versa. Recently, combined instruments became commercially available, where dispersive Raman micro-spectroscopy can easily be combined with ATR infrared microscopy. ATR (attenuated total reflection) is a technique where the sample is pressed with an infrared-transparent crystal and the interface between the sample and the crystal is studied by infrared-spectroscopy. By using this approach, it is possible to obtain infrared spectra with good spatial resolution.

In some laboratories, specific mechanical properties of substances are studied, for instance to evaluate the strength of fibres. These benches can be positioned under a Raman microscope, and while performing the stress test, Raman spectra are recorded. Thus it is possible to study the influence of the mechanical stress on the molecular level. Also, heating or cooling stages are sometimes used to study the thermal properties of products (e.g. differential thermal analysis, DTA) and the molecular changes that occur during heating or cooling.

Sometimes it is very useful to combine molecular spectroscopic measurements with the determination of the elemental composition. In this case, it is possible to use a combined Raman spectrometer with a scanning electron microscope (SEM), which is rare but commercially available. The measurements are not performed simultaneously: as the point is measured with SEM, the stage is translated and the Raman measurement is performed. One disadvantage of this approach is that the spot sizes for both techniques are different. Moreover, as during SEM measurements the sample is bombarded with electrons, it should be electrically conductive, and sometimes it is necessary to coat the sample – which may hamper the Raman measurements. In practice, for most measurements it is possible to record the Raman spectra and then move the sample to another laboratory, where the SEM-instrument is available.

Also the combination between Raman spectroscopy and X-ray fluorescence spectrometry has been made, but to our knowledge there is no combined Raman-XRF instrument commercially available on the market. Mostly, the measurements are made in sequence, where XRF usually is the fastest technique, which can be used for getting an overview of the elemental composition of the sample, and afterwards the Raman measurements are made.

A last combination which is used, is the combination of Raman spectroscopy with laser-induced breakdown spectroscopy. In this approach, first a low power laser is used to record the Raman spectrum and subsequently, using the same optics, a high power laser is used to ablate part of the material. As a consequence, a plume is formed where the sample is partially atomised and ionised and therefore the sample emits light that is specific for its composition: mostly elemental information is obtained, but some information on functional groups can be found. These results can readily be combined with the results from the Raman measurements, as both measurements are performed on exactly the same location.

It is clear that in the near future further combinations will be developed and that Raman spectroscopy will continue to provide useful information for scientists, worldwide.

5.9 Summary

In this chapter, some general aspects of daily practices in Raman spectroscopy have been described. We have discussed the way to calibrate a Raman spectrometer. Another aspect of today's practices in Raman spectroscopy is the use of digital filters for processing the results and the use of spectral searching algorithms. In this chapter we have also discussed some general aspects of Raman spectral interpretation of inorganic and organic molecules and paid attention to some pitfalls concerning quantitative aspects of Raman spectroscopy. Finally, we have also discussed the possibilities for Raman mapping and imaging.

Further Reading

Graselli J.G., Bulkin B.J. (eds), *Analytical Raman Spectroscopy*, John Wiley & Sons, Inc., New York, 1991.

Lewis, I.R. and Edwards, H.G.M. (eds), *Handbook of Raman Spectroscopy – From the Research Laboratory to the Process Line*, Marcel Dekker, New York, 2001.

Lin-Vien, D., Colthup, N.B., Fateley, W.G., Grasselli, J.G., *The Handbook of Infrared and Raman Characteristic Frequencies of Organic Molecules*, Academic Press, San Diego, 1991.

Long, D.A., *Raman Spectroscopy*, McGraw-Hill, Maidenhead, UK, 1977.

Pelletier M.J. (ed.), *Analytical Applications of Raman Spectroscopy*, Blackwell Science, Malden, 1999.

Socrates, G., *Infrared and Raman Characteristic Group Frequencies – Tables and Charts*, 3rd Edition, John Wiley & Sons Ltd, Chichester 2001.

Responses to Questions

QUESTION 1.1

1st term : 10^{-40} CV^{-1}m$^2 \cdot 3 \cdot 10^6$ Vm$^{-1} = 3 \cdot 10^{-33}$ C.m

2nd term : $(1/2) \cdot 10^{-50}$ CV^{-2}m$^3 \cdot (3 \cdot 10^6$ Vm$^{-1})^2 = 4.5 \cdot 10^{-37}$ C.m

3rd term : $(1/6) \cdot 10^{-60}$ CV^{-3}m$^4 \cdot (3 \cdot 10^6$ Vm$^{-1})^3 = 4.5 \cdot 10^{-41}$ C.m

---

QUESTION 1.2

785 nm corresponds to $785 \cdot 10^{-9}$ m or $785 \cdot 10^{-7}$ C.m. This equals, in absolute wavenumbers 12 738.85 cm^{-1}.

A Raman shift of 352 cm^{-1} corresponds with the absolute wavenumbers 13090,85 and 12 386.85 cm^{-1} for Anti-Stokes and Stokes radiation, respectively. This corresponds with 763.89 nm and 807.31 nm, respectively.

Practical Raman Spectroscopy: An Introduction, First Edition. Peter Vandenabeele.
© 2013 John Wiley & Sons, Ltd. Published 2013 by John Wiley & Sons, Ltd.

QUESTION 1.3

The CO_2 molecule has 4 normal vibrations: the symmetrical stretch and the asymmetrical stretch, and two bending vibrations which are perpendicular to each other. The two bending vibrations are degenerated: they correspond with the same energy transitions. During symmetrical stretching, the polarisability of the molecule changes, but not the dipole, so this vibration is Raman active, but not infrared active. During the asymmetrical stretching, the dipole moment changes, but not the over-all polarisability, so this vibration is infrared active, but not Raman active. During the bending vibration, there is a change in dipole of the molecule, but no change in polarisability, so this vibration is infrared active, but not Raman active. Note that the rule of mutual exclusion for centrosymmetrical molecules applies here: vibrational modes that are Raman active are infrared inactive and vice versa.

QUESTION 1.4

Every Raman spectrum (Figure 1.1) has a very intense band at $0 \, cm^{-1}$: the *Rayleigh* line, caused by *elastic* scattering of electromagnetic radiation. Normally the intensity of this line is suppressed by using appropriate filters.

At positive wavenumbers, *Stokes* Raman bands are observed, while at negative wavenumbers, one can find *anti-Stokes* Raman, bands. We know that *Stokes* Raman scattering is more intense than *anti-Stokes* scattering, as a consequence of the *Boltzmann* distribution.

Stokes and anti-Stokes bands are positioned symmetrically to the Rayleigh line. The observed Raman bands correspond with transitions for which there is a change in *polarisability* of the molecule. By using group theory, it is possible to study whether certain transitions will give rise to a Raman band or not.

The Raman band position is dependent on the *force constant* of the bond, as well as of the *reduced mass* of the constituting atoms. As a consequence, *group frequencies* can be defined. Bonds with relatively heavy elements (typically inorganic materials, such as metal oxides) give rise to Raman bands at relatively *low* wavenumbers. Moreover, crystalline materials can undergo lattice vibrations: these can hardly be considered as intramolecular vibrations, but are merely vibrations of larger units relative to each other. The corresponding Raman bands are observed at *low* wavenumbers and are strongly dependent of the local symmetry. Amorphous materials in general give rise to *broader* Raman bands, compared to crystalline materials.

QUESTION 4.1

When using a Nd:YAG laser with a wavelength of 1064 nm, the Stokes band at $3000\,cm^{-1}$ is detected at 1563 nm. However, when using a HeNe laser of 632.8 nm, the same band is detected at 781 nm.

QUESTION 4.2

Around 785 nm, a shift of $1\,cm^{-1}$ corresponds to 0.062 nm, around 632.5, the same shift corresponds with 0.040 nm, whereas for a 413.1 nm laser, this corresponds with 0.017 nm.

QUESTION 4.3

A 10% filter has an OD of $-\log(0.1) = 1$
A 25% filter has an OD of $-\log(0.25) = 0.60$
A filter with an OD of 0.30 is a $10^{-0.30} = 0.50$ or 50% filter

QUESTION 4.4

Long pass filters only allow radiation with a longer wavelength to pass and block radiation with a lower wavelength. As a consequence, anti-Stokes radiation is blocked and when using a dielectric filter it is not possible to record anti-Stokes spectra.

QUESTION 5.1

99-99-98-97-96-97-97-96-98-100-104-106-113-126

Bibliography

Banwell, C.N., *Fundamentals of Molecular Spectroscopy*, 3rd Edition, McGraw-Hill, London, UK, 1983.

Laserna, J.J (ed.), *Modern Techniques in Raman Spectroscopy*, John Wiley & Sons, Ltd, Chichester, UK, 1996.

Ferraro, J.R., Nakamoto, K., Brown, C.W., *Introductory Raman Spectroscopy*, 2nd Edition, Academic Press, Amsterdam 2003.

Gardiner, D.J. and Graves, P.R. (eds), *Practical Raman Spectroscopy*, Springer, New York, 1988.

Graselli J.G., Bulkin B.J. (eds), *Analytical Raman Spectroscopy*, John Wiley & Sons, Inc., New York, 1991.

Lewis, I.R. and Edwards, H.H.M. (eds), *Handbook of Raman spectroscopy – From the Research Laboratory to the Process Line*, Marcel Dekker, Inc., New York, 2001.

Lin-Vien, D., Colthup, N.B., Fateley, W.G., Grasselli, J.G., *The Handbook of Infrared and Raman Characteristic Frequencies of Organic Molecules*, Academic Press, San Diego, 1991.

Long, D.A., *Raman Spectroscopy*, McGraw-Hill, Maidenhead, UK, 1977.

McCreery, R.L. *Raman Spectroscopy for Chemical Analysis*, John Wiley & Sons, Inc., New York, 2000.

Pelletier M.J. (ed.), *Analytical Applications of Raman Spectroscopy*, Blackwell Science, Malden, 1999.

Smith, E. and Dent, G., *Modern Raman Spectroscopy: A Practical Approach*, John Wiley & Sons Ltd, Chichester, UK, 2005.

Practical Raman Spectroscopy: An Introduction, First Edition. Peter Vandenabeele.
© 2013 John Wiley & Sons, Ltd. Published 2013 by John Wiley & Sons, Ltd.

Glossary of Terms

Anisotropy   Term used to indicate that optical effects are dependent on the orientation in space.

Anti-Stokes scattering   Raman scattering where the scattered light has a shorter wavelength (higher energy) than the incident beam.

Confocal measurements   Approach where a confocal pinhole is positioned in the lightpath, so that only information of a specific depth (axial resolution) is obtained.

Fluorescence   Scattering of radiation caused by electronic transitions.

Polarisability   Quantity expressing how easily electrons in a molecule can be moved to induce a dipole.

Quantum efficiency   Quantity used to express the sensitivity of a detector. Expresses how many photons correspond with a detected electron or electron–hole pair.

Rayleigh scattering   Elastic scattering where the scattered light has the same wavelength as the incident beam.

Resonance Raman effect   Raman effect that is enhanced as the virtual energy level coincides with an electronic energy level.

Practical Raman Spectroscopy: An Introduction, First Edition. Peter Vandenabeele.
© 2013 John Wiley & Sons, Ltd. Published 2013 by John Wiley & Sons, Ltd.

Selection rule   Expresses whether certain transitions are allowed
  or not and whether they give rise to radiation.
Stokes scattering   Raman scattering where the scattered light has
  a longer wavelength (lower energy) than the incident beam.

Index

Note: Page numbers in *italic* refer to figures and tables.

Practical Raman Spectroscopy: An Introduction, First Edition. Peter Vandenabeele.
© 2013 John Wiley & Sons, Ltd. Published 2013 by John Wiley & Sons, Ltd.

SI Units and Physical Constants

SI Units

The SI system of units is generally used throughout this book. It should be noted, however, that according to present practice, there are some exceptions to this, for example, wavenumber (cm^{-1} and ionization energy (eV).

Base SI units and physical quantities

Quantity	Symbol	SI unit	Symbol
length	l	metre	m
mass	m	kilogram	kg
time	t	second	s
electric current	I	ampere	A
thermodynamic temperature	T	kelvin	K
amount of substance	n	mole	mol
luminous intensity	I_v	candela	cd

Practical Raman Spectroscopy: An Introduction, First Edition. Peter Vandenabeele.
© 2013 John Wiley & Sons, Ltd. Published 2013 by John Wiley & Sons, Ltd.

Prefixes used for SI units

Factor	Prefix	Symbol
10^{21}	zetta	Z
10^{18}	exa	E
10^{15}	peta	P
10^{12}	tera	T
10^{9}	giga	G
10^{6}	mega	M
10^{3}	kilo	k
10^{2}	hecto	h
10	deca	da
10^{-1}	deci	d
10^{-2}	centi	c
10^{-3}	milli	m
10^{-6}	micro	μ
10^{-9}	nano	n
10^{-12}	pico	p
10^{-15}	femto	f
10^{-18}	atto	a
10^{-21}	zepto	z

SI Units and Physical Constants

Derived SI units with special names and symbols

Physical quantity	SI unit		Expression in terms of base or derived SI units
	Name	Symbol	
frequency	hertz	Hz	$1\,\text{Hz} = 1\,\text{s}^{-1}$
force	newton	N	$1\,\text{N} = 1\,\text{kg m s}^{-2}$
pressure; stress	pascal	Pa	$1\,\text{Pa} = 1\,\text{N m}^{-2}$
energy; work; quantity of heat	joule	J	$1\,\text{J} = 1\,\text{Nm}$
power	watt	W	$1\,\text{W} = 1\,\text{J s}^{-1}$
electric charge; quantity of electricity	coulomb	C	$1\,\text{C} = 1\,\text{A s}$
electric potential; potential difference; electromotive force; tension	volt	V	$1\,\text{V} = 1\,\text{JC}^{-1}$
electric capacitance	farad	F	$1\,\text{F} = 1\,\text{C V}^{-1}$
electric resistance	ohm	Ω	$1\,\Omega = 1\,\text{V}^{-1}$
electric conductance	siemens	S	$1\,\text{S} = 1\,\Omega^{-1}$
magnetic flux; flux of magnetic induction	weber	Wb	$1\,\text{Wb} = 1\,\text{V s}$
magnetic flux density; magnetic induction	tesla	T	$1\,\text{T} = 1\,\text{Wb m}^{-2}$
inductance	henry	H	$1\,\text{H} = 1\,\text{Wb A}^{-1}$

(continued overleaf)

Derived SI units with special names and symbols (*continued*)

Physical quantity	SI unit		Expression in terms of base or derived SI units
	Name	Symbol	
Celsius temperature	degree Celsius	°C	$1°C = 1\,K$
luminous flux	lumen	lm	$1\,lm = 1\,cd\,sr$
illuminance	lux	lx	$1\,lx = 1\,lm\,m^{-2}$
activity (of a radionuclide)	becquerel	Bq	$1\,Bq = 1\,s^{-1}$
absorbed dose; specific energy	gray	Gy	$1\,Gy = 1\,J\,kg^{-1}$
dose equivalent	sievert	Sv	$1\,sv = 1\,J\,kg^{-1}$
plane angle	radian	rad	1^a
solid angle	steradian	sr	1^a

[a] rad and sr may be included or omitted in expressions for the derived units.

Physical Constants

Recommended values of selected physical constants[a]

Constant	Symbol	Value
acceleration of free fall (acceleration due to gravity)	g_n	$9.806\,65\,m\,s^{-2}$ [b]
atomic mass constant (unified atomic mass unit)	m_u	$1.666\,540\,2(10) \times 10^{-27}\,kg$

Recommended values of selected physical constants[a]

Constant	Symbol	Value
Avogadro constant	L, N_A	$6.022\ 136\ 7(36) \times 10^{23}$ mol^{-1}
Boltzmann constant	k	$1.380\ 658(12) \times 10^{-23}\ \text{JK}^{-1}$
electron specific charge (charge-to-mass ratio)	$-e/m_e$	$-1.758\ 819 \times 10^{11}\ \text{C kg}^{-1}$
electron charge (elementary charge)	e	$1.602\ 177\ 33(49) \times 10^{-19}\ \text{C}$
Faraday constant	F	$9.648\ 530\ 9(29) \times 10^4\ \text{C}$ mol^{-1}
ice-point temperature	T_{ice}	$273.15\ \text{K}^b$
molar gas constant	R	$8.314\ 510(70)\ \text{J K}^{-1}\ \text{mol}^{-1}$
molar volume of ideal gas (at 273.15 K and 101 325 Pa)	V_m	$22.414\ 10(19) \times 10^{-3}\ \text{m}^3$ mol^{-1}
Planck constant	h	$6.626\ 075\ 5(40) \times 10^{-34}\ \text{J s}$
standard atmosphere	atm	$101\ 325\ \text{Pa}^b$
speed of light in vacuum	c	$2.997\ 924\ 58 \times 10^8\ \text{m s}^{-1\ b}$

[a]Data are presented in their full precision, although often no more than the first four or five significant digits are used; figures in parentheses represent the standard deviation uncertainty in the least significant digits.
[b]Exactly defined values.